Erhard Bühler
Dieter Franke

Topics in Identification and Distributed Parameter Systems

Advances in Control Systems and Signal Processing

Editor: Irmfried Hartmann

Volume 1

Volume 1: Erhard Bühler and Dieter Franke
Topics in Identification and Distributed Parameter Systems

Erhard Bühler and Dieter Franke

Topics in Identification and Distributed Parameter Systems

With 42 Figures

Friedr. Vieweg & Sohn Braunschweig/Wiesbaden

CIP-Kurztitelaufnahme der Deutschen Bibliothek

Bühler, Erhard:
Topics in identification and distributed parameter systems/Erhard Bühler and Dieter Franke. – Braunschweig, Wiesbaden: Vieweg, 1980.
(Advances in control systems and signal processing; Vol. 1)
ISBN 3-528-08469-3

NE: Franke, Dieter:

Editor:

Dr.-Ing. I. Hartmann
o. Prof. für Regelungstechnik und Systemdynamik
Technische Universität Berlin
Hardenbergstraße 29c
1000 Berlin 12, West Germany

Authors:

Dr.-Ing. E. Bühler
Spiegelberg 1
7759 Immenstead, West Germany

Dr.-Ing. D. Franke
o. Prof. für Regelungstechnik
Hochschule der Bundeswehr Hamburg
Holstenhofweg 85
2000 Hamburg 70, West Germany

Druck: fotokop, Darmstadt
Buchbinderische Verarbeitung: Paul Junghans, Darmstadt
Printed in Germany

ISBN 3-528-08469-3

Preface

This bookseries aimes to report new developments on parts of control systems and signal processing at a high level, without a restriction in the size of contribution, if (it is) necessary. Basic knowledge is needed.

Each volume deals with a special theme. The following types of contributions will be taken into consideration:

- coherent representation on single topics with new results,
- new improved expositions or new applications in a known field of control systems and signal processing,
- practical applications of theoretic methods,
- a survey of published articles with deep insight.

Seminar work outs or themes from meetings with exeptional interest could belong to this series, if they satisfy the conditions.

The manuscript is reproduced by photographic process and therefore it must be typed with extreme care. The author can get instruction.

The manuscript is sent to Prof. Dr. I. Hartmann, Institut für Regelungstechnik, Technische Universität Berlin, Hardenbergstraße 29c, 1000 Berlin 12, West Germany.

Irmfried Hartmann

Berlin, January 1980

Contents

Contribution I

Control of bilinear distributed parameter systems
Dieter Franke

Contribution II

Contribution I

Control of Bilinear Distributed Parameter Systems

Dieter Franke

Contribution I

Control of bilinear distributed parameter systems

Dieter Franke

Notations

$\underline{A}$ linear partial differential matrix operator with respect to $\underline{z}$

$\tilde{\underline{A}}$ system matrix of linear lumped subsystem

$\underline{B}_i$ linear partial differential matrix operators with respect to $\underline{z}$

$\tilde{\underline{B}}$ input matrix of linear lumped subsystem

$\tilde{\underline{C}}$ output matrix of linear lumped subsystem

$\underline{C}(\underline{z})$ matrix depending on $\underline{z}\in\Omega$

$\underline{D}(\underline{z})$ matrix depending on $\underline{z}\in\Gamma$

D region in Euclidean space

$\underline{E}$ matrix integral output operator

$\underline{f}$ vector-valued nonlinear function

$\underline{g}$ vector-valued nonlinear function

$\underline{G}(t-\tau,\underline{z},\underline{\zeta})$ Green's matrix

$g(t-\tau,z,\zeta)$ scalar Green's function

$g_i(t-\tau,z)$ kernels derived from the Green's function

I, $\underline{I}$ identity operator, identity matrix

$\underline{I}_\Omega$, $\underline{I}_\Gamma$, $\underline{I}_o$ linear matrix integral operators

J cost functional

K, k scalar gain

$\underline{K}$, $\underline{K}_1$, $\underline{K}_2$ gain matrices

l(t) time varying length, moving boundary

m order of lumped subsystem

m_i, M_i bounds on control variable $u_i(t)$

n dimension of state vector of distributed system

p dimension of total control vector $\underline{u}(t)$

p_1 dimension of control vector $\underline{u}^{(1)}(t)$ acting in Ω

$p-p_1$ dimension of control vector $\underline{u}^{(2)}(t)$ acting on Γ

P point from region D

$\underline{P}$ positive definite matrix

q dimension of output vector $\underline{y}(t)$

Q point from region D

$\underline{Q}$ symmetric matrix

$\underline{R}$ linear partial differential matrix operator with respect to $\underline{z}$

S admissible set for input vector $\underline{u}(t)$

t time

T finite terminal time

$\underline{u}(t)$ control vector, input vector

$\underline{u}_\Omega(t,\underline{z})$ vector-valued forcing function in domain Ω

$\underline{u}_\Gamma(t,\underline{z})$ vector-valued forcing function on boundary Γ

U domain of negative definite $\underline{Q}$

V Ljapunow-functional

$w(z)$ pre-assigned terminal distribution

$\underline{w}$ reference input vector

$\underline{x}(t,\underline{z})$ state vector of distributed parameter system

$x_i^*(t)$ generalized Fourier-coefficients

$X_T(\underline{x}_o)$ set of states reachable from $\underline{x}_o(\underline{z})$ in time T

$\hat{X}(\underline{x}_o)$ reachable set from $\underline{x}_o(\underline{z})$

$\underline{y}(t)$ output vector

$\underline{y}_e(t)$ vector-valued error

$\underline{z}$ vector of spatial coordinates

Γ boundary of domain Ω

$\delta(z,z_M)$ spatial dirac function at $z = z_M$

$\underline{\zeta}$ vector of spatial coordinates

$\underline{\eta}(t,\underline{z})$ total state vector of mixed lumped and distributed parameter system

Θ domain of negative semidefinite $\dot{V}$

λ_i Eigenvalues

Λ domain of asymptotic stability

μ dimension of space vector $\underline{z}$

$\underline{\xi}(t)$ state vector of lumped subsystem

$\Pi(P,\underline{u})$ Butkovskiy's function

$\rho(.,.)$ metric

τ time

$\{\varphi_i(z)\}$ complete set of functions, orthonormal eigenfunctions

$\underline{\Phi}(t)$ transition matrix

$\underline{\Psi}(P)$ adjoint vector

Ω fixed spatial domain in μ-dimensional Euclidean space

Δ denotes deviation from an equilibrium state

o (subscript) denotes initial state

s (subcript) denotes steady state

T (superscript) denotes transpose

$||.||$ L_2 - norm

1. Introduction

In recent years, considerable interest has been focussed on a special class of nonlinear dynamical systems, usually called *bilinear systems*. They are linear in state and linear in control, but not jointly linear in both. Hence, products of state and control variables are a characteristic feature of this type of systems [1]-[4], [6], [7], [10]-[15].

As far as ordinary differential equations are considered, the state equations of a bilinear system therefore take the form

$$\dot{\underline{x}}(t) = \underline{A}\ \underline{x}(t) + \sum_{i=1}^{p} u_i(t)\underline{B}_i\underline{x}(t) + \underline{C}\,\underline{u}(t), \tag{1.1}$$

where $\underline{x}$ is an n-dimensional state vector, $\underline{u}$ is a p-dimensional control vector, $\underline{A}$, $\underline{B}_i$, i=1, ..., p, and $\underline{C}$ are suitably-typed constant matrices. Of course, the linear case is contained in Eq.(1.1) for $\underline{B}_i = \underline{O}$, i=1, ..., p.

Why this engagement in bilinear systems?

First, a great variety of control processes, not only in engineering, but also in socio-economics, biological systems and other fields, can be modelled by bilinear state equations. Bilinear structures seem to be quite common in nature due to their highly adaptive character [12].

Therefore, by learning from nature, bilinear control modes may be utilized to the design of more effective control devices than available by linear methods. Indeed, by rewriting Eq.(1.1) in the form

$$\dot{\underline{x}}(t) = [\underline{A} + \sum_{i=1}^{p} u_i(t)\underline{B}_i]\underline{x}(t) + \underline{C}\,\underline{u}(t), \tag{1.2}$$

it can be seen that the input $\underline{u}$ acts as a *parametric* control, in addition to ordinary linear control. Obviously, parametric (or multiplicative) control action is an effective tool for changing the system matrix $\underline{A}$, and for this reason bilinear systems are often referred to as *variable structure systems*.

Apart from applicative interest, bilinear systems are an "appealing class of 'nearly linear' systems" [4] also in theory. Due to the special type of nonlinearity, linear techniques may be employed to some extent. If, for example, the control is piecewise constant, then the bilinear system is piecewise linear. The set of equilibrium states subject to constant input can be obtained by the linear technique of matrix inversion.

Questions of mathematical modelling, identification, structural properties (stability, reachability, controllability and observability) and optimization are of greatest interest in the current research in this field. An excellent survey of the state of the art has been given by Bruno, Di Pillo and Koch [4].

One of the most significant benefits of multiplicative control is the improvement of controllability in the presence of constraints on the control. Mohler [12] has shown that stable linear systems with *bounded* control are not completely controllable even if Kalman's conditions [44] are satisfied. The close relationship between stability and controllability has been exploited by Mohler to provide sufficient conditions for complete controllability of bilinear systems.

Moreover, as a consequence of improved controllability, bilinear systems exhibit far better performance than linear systems, when optimal control methods are applied [2], [7], [12].

Whereas the theory of those bilinear systems, modelled by ordinary state equations, is now established to a great extent, there are only a few papers concerned with bilinear distributed parameter systems, described by partial differential equations [33]-[38]. These contributions are very special, and therefore it seems to the author that there is a need of a very fundamental entering into the field. Moreover, the exploration of bilinear distributed parameter systems is expected to be of significant practical interest, as most systems in engineering and elsewhere are distributed in space. Generally speaking, a linear distributed system becomes a bilinear one, whenever some coefficients in the underlying state equations can be varied within given bounds. Parametric control action in distributed parameter systems turns out not only to be easy to implement in many situations, but also to be more effective than linear control.

Of course, any distributed system may be approximated by a lumped system of sufficiently high order. But there is a number of specific methods for distributed parameter systems, which require late approximation or even make any approximation dispensible, such as Ljapunow's direct method ([18], [24],[56]) or certain methods of functional analysis [53]. In these situations lumped approximation would destroy the character of the problem.

The material of this contribution is to be understood as a first approach to the very complex field of bilinear distributed parameter systems. Therefore only continuous-time, deterministic and time invariant systems are considered. With regard to practical implementation vector-valued lumped input and lumped output are preferably assumed. For the ease of presentation, scalar partial differential state equations are underlying in some of the proposed methods. In many situations, the actuator for a bilinear distributed parameter plant will have lumped parameters. Therefore, special attention will be focussed on mixed lumped and distributed control processes.

To fix ideas, chapter 2 starts with some very simple practical examples of bilinear distributed plants, such as continuous flow problems and heat exchange processes. The same examples are used throughout the whole text to illustrate the methods presented. Systems with a moving boundary, with boundary *location* acting as a control variable, will be detected to be of inherent bilinear character by means of a simple transformation. This class of systems seems to be quite common in population dynamics. By abstracting from the concrete examples, the general state equations of bilinear distributed systems are given in differential and in integral form.

Chapter 3 presents some very basic remarks on structural properties. Therefore, the equilibrium set is determined via Green's functions methods, then stability analysis of a class of bilinear systems is carried out by means of Ljapunow's direct method. Reachability and controllability are briefly discussed, and a sufficient condition of reachability is given, basing on the method of moments. Here, a wide field for future research opens.

As in the lumped parameter case, linear feedback of a bilinear plant leads to the class of so-called *quadratic-in-the-state* dynamical systems. The state equations of mixed lumped and distributed parameter systems of this type are derived in chapter 4. In the context of feedback control, another class of bilinear distributed systems is of importance: variable structure control of linear plants, considered by Becker [1], [2] as a special class of bilinear systems in the lumped parameter case. The state equations of both, quadratic-in-the-state systems and variable structure control systems, will serve as a basis for the design of control methods to be proposed in chapters 5 and 6.

Ljapunow's direct method turns out to be an efficient tool for the design of controllers for bilinear distributed plants to be proposed in chapter 5. Both, linear and nonlinear feedback laws are constructed, with special attention to mixed lumped and distributed over-all systems with bounded control variable. The lumped subsystem is allowed to be of arbitrary order and the distributed subsystem need not be approximated.

The design of optimal control systems is the object of chapter 6. Butkovskiy's maximum principle proves to be a comprehensive method for the optimal control of a wide class of nonlinear distributed systems, basing on integral state equations. This makes it also an appropriate tool for treating bilinear systems. First, a class of optimal bilinear boundary open-loop control problems is considered. This demonstrates that even in the presence of very restricted reachability properties favourable performance can be achieved. Finally, an optimal switching strategy for mixed lumped and distributed closed-loop systems with variable structure is computed, using again Butkovskiy's theory.

The whole material of this contribution has been tried to be presented not too mathematically, sometimes in a merely intuitive way. The author, engineer himself, hopes that this style will motivate the reader, from whatever discipline, to further research and applications in the significant field of bilinear distributed parameter systems.

2. Mathematical models of bilinear distributed parameter processes

2.1. Processes with parametric control action

The objective of this section is to provide some very basic engineering examples of distributed parameter processes which may motivate the application of multiplicative control actions. The examples try to illustrate that from the view of practical implementation multiplicative control variables are in many situations more natural and more effective than linear control.

Example 2.1:

Consider the simple process of transportation of some material, outlined in Fig. 2.1. The velocity of the conveyor-belt, $u_1(t)$, can be assumed to be independent of the spatial variable z, and therefore the well-known (normed) equation

$$\frac{\partial x(t,z)}{\partial t} + u_1(t)\,\frac{\partial x(t,z)}{\partial z} = u_2(t,z),\; 0 \leq z \leq 1, \qquad (2.1)$$

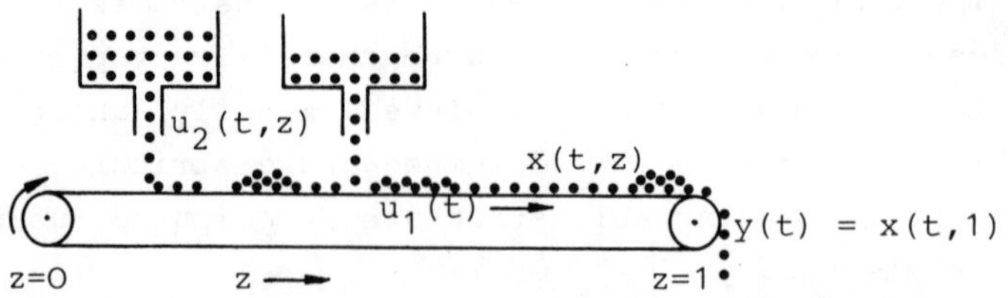

Fig.2.1 : Illustration of a conveyor equipment

is valid for the spatial distribution $x(t,z)$ of the material. $u_2(t,z)$ is the feed which may be lumped or distributed.

Boundary condition

$$x(t,0) = 0 \tag{2.2}$$

and initial distribution

$$x(0,z) = x_0(z) \tag{2.3}$$

complete the model.

If it is required to maintain a specified value of the output

$$y(t) = x(t,1), \tag{2.4}$$

then $u_1(t)$ will be an appropriate control variable. In addition to simple implementation, this multiplicative mode may considerably reduce the time lag involved with purely linear control.

Example 2.2:

Figure 2.2 outlines a schematic diagram of a continuous heating or cooling processing zone. Generally speaking, material moves through the zone where a certain field is acting on some quality of the material.

By considerably simplifying the underlying equations, one arrives at the form [17]

$$\frac{\partial x(t,z)}{\partial t} + u_1(t)\frac{\partial x(t,z)}{\partial z} + u_2(t,z)x(t,z) = u_2(t,z)u_3(t,z), \quad 0 \leq z \leq 1, \tag{2.5}$$

where $x(t,z)$ is the temperature inside the material, $u_1(t)$ is the velocity of transportation, $u_2(t,z)$ is a normed heat transfer coefficient, and $u_3(t,z)$ is the field of the acting temperature. Again a boundary condition

$$x(t,0) = u_4(t) \tag{2.6}$$

at the entry and the initial distribution

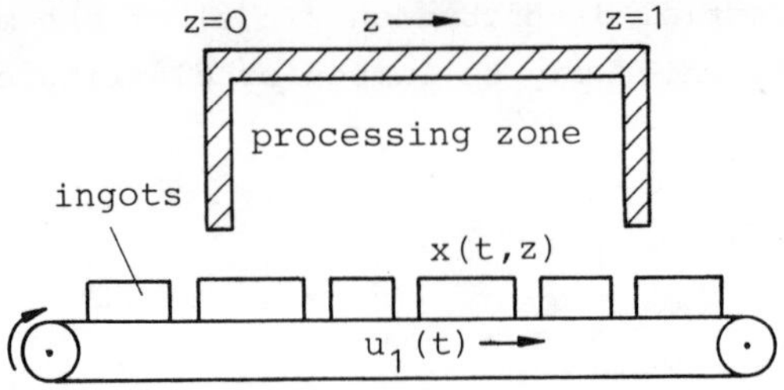

Fig.2.2 : Diagram of a continuous processing zone

$$x(0,z) = x_o(z) \quad (2.7)$$

may be specified.

Obviously, the temperature of the material at the outlet,

$$y(t) = x(t,1), \quad (2.8)$$

depends on the control variables u_1 up to u_4. By means of the abbreviation

$$\tilde{u}_3(t,z) := u_2(t,z)u_3(t,z), \quad (2.9)$$

the model takes a bilinear nature, with two linear control modes, $\tilde{u}_3$ and u_4, and two multiplicative modes, u_1 and u_2.

In case of very small $u_1(t)$ or even $u_1(t) \equiv 0$ during some intervals of time, axial heat conduction should not be neglected which gives rise to the additional term $- a \cdot \partial^2 x/\partial z^2$ on the left hand side of Eq.(2.5).

In examples 2.1 and 2.2 the candidates of multiplicative control occur as coefficients in the underlying partial differential equation. Example 2.3 is an example of multiplicative *boundary* control.

Example 2.3:

In power plants there arises the permanent necessity of cooling aggregates such as nuclear reactors, transformers or generators. In many situations the thermal nature of these processes may be schematically diagrammed as shown in Fig.2.3

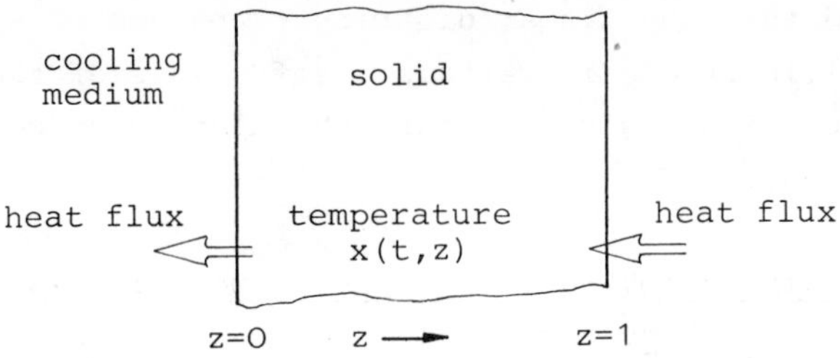

Fig.2.3 : Illustration of a one-dimensional cooling problem

and modelled by the well-known heat equation with appropriate boundary conditions. A one-dimensional version may be given by

$$\frac{\partial x(t,z)}{\partial t} - a\,\frac{\partial^2 x(t,z)}{\partial z^2} = u_1(t,z), \quad 0 \leq z \leq 1, \quad a>0, \tag{2.10}$$

with boundary conditions

$$\left.\frac{\partial x(t,z)}{\partial z}\right|_{z=0} = u_2(t)\,x(t,0), \tag{2.11}$$

$$\left.\frac{\partial x(t,z)}{\partial z}\right|_{z=1} = u_3(t), \tag{2.12}$$

and initial condition

$$x(0,z) = x_o(z). \tag{2.13}$$

Here, $x(t,z)$ is the temperature inside the aggregate to be cooled from the boundary $z=0$. Hence, $u_2(t)$ is a normed heat transfer coefficient to be varied via coolant *flow rate*. The coolant *temperature* has been set equal to zero for simplification. $u_1(t,z)$ and $u_3(t)$ are inputs caused by heat generation of the aggregate. Of course, $u_2(t)$ is an extremely important boundary control variable to maintain a specified temperature. This may again illustrate the benefit of a multiplicative mode.

A number of more complex distributed parameter processes with inherent bilinear structure such as heat exchangers or drying plants are basing on these simple examples. In general, it can be concluded that any linear distributed parameter system can be made a bilinear one if certain coefficients of the model may be varied within given bounds, thus acting as control variables.

2.2 Systems with a moving boundary - a special class of "nearly linear" systems

In recent years, considerable interest has been focussed on a class of distributed parameter systems usually referred to as *moving boundary problems* [50], [59]. They arise in some technical applications such as melting and solidification in metallurgical plants. In certain mechanical constructions such as crane bridges or industrial roboter arms the effective *length* of elastic beams may be required to be specified functions of time thus giving rise to moving boundary problems (see example 2.5).

Systems with moving boundaries are thought to be quite common in nature. The population of a species will generally be a function of time and of spatial variables. The boundary of the living space will usually vary rather than to be fixed, due to time varying conditions and environment properties.

Example 2.4:

Hoppenstaedt [43] has proposed a diffusion type mathematical model for the fish population distributed in a habitat bounded on one side (z=0) by a breeding ground and on the other (z=1) by an unfavourable environment. The essential part of the model may briefly be reported as follows:

$$\frac{\partial x(t,z)}{\partial t} = \sigma^2 \frac{\partial^2 x(t,z)}{\partial z^2} - \alpha x(t,z) - u_1(t,z), \quad 0 \leq z \leq 1, \qquad (2.14)$$

with boundary conditions

$$\left. \frac{\partial x(t,z)}{\partial z} \right|_{z=0} = - u_2(t), \qquad (2.15)$$

$$\left. \frac{\partial x(t,z)}{\partial z} \right|_{z=1} = 0 \qquad (2.16)$$

and initial condition

$$x(0,z) = x_o(z). \tag{2.17}$$

Here σ^2 measures the dispersal rate of the fish, α is the combined removal rate, u_1 is the harvesting rate density and u_2 the flux into adult habitat from breeding ground.

So far this is a fixed-domain linear system. Suppose now the fixed boundary z=1 to be replaced by a moving boundary z=l(t)>0. The function l(t) may either be varied by man or may vary due to certain properties of the environment. l(t) is assumed to have a continuous derivative $\dot{l}(t)$.

The relation

$$\dot{l}(t) = \left.\frac{\partial u_3(t,z)}{\partial z}\right|_{z=l(t)} \tag{2.18}$$

may serve as an example. Here u_3 is an environment property favourable to an extension of the population.

By means of the simple transformation

$$\zeta = z/l(t) \tag{2.19}$$

proposed by Wang [24], the new spatial variable ζ may be introduced which will make the transformed system a fixed-domain system. In detail the following set is obtained from Eqs.(2.14) - (2.17):

$$\frac{\partial x}{\partial t} = \zeta \frac{\dot{l}}{l} \frac{\partial x}{\partial \zeta} + \frac{\sigma^2}{l^2} \frac{\partial^2 x}{\partial \zeta^2} - \alpha x - u_1, \quad 0 \leq \zeta \leq 1 \tag{2.20}$$

with boundary conditions

$$\left.\frac{\partial x}{\partial \zeta}\right|_{\zeta=0} = - l u_2, \tag{2.21}$$

$$\left.\frac{\partial x}{\partial \zeta}\right|_{\zeta=1} = 0 \tag{2.22}$$

and initial condition

$$x(0,\zeta) = x_o[\zeta l(0)]. \tag{2.23}$$

Obviously, the transformed system involves essentially a *bilinear part* with linear controls $u_1(t)$ and $\tilde{u}_2(t) = -l(t)u_2(t)$ and multiplicative controls

$$\tilde{u}_3(t) = \dot{l}(t)/l(t), \tag{2.24}$$

$$\tilde{u}_4(t) = \sigma^2/l^2(t). \tag{2.25}$$

The structure of this biological system is outlined in Fig.2.4. The inherent bilinear structure of moving boundary problems seems to have not been stressed in the past.

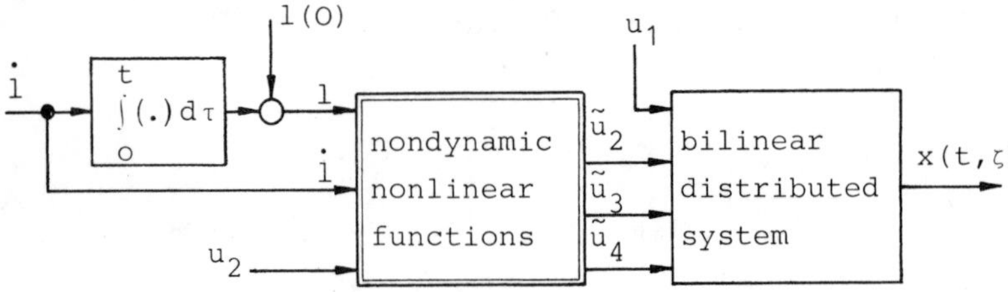

Fig.2.4: Structure of population dynamics with moving boundary $l(t)$

Example 2.5:

As an engineering example of a moving boundary problem consider the transverse vibrations of an elastic beam with variable effective length $l(t)$, as outlined in Fig.2.5. Again the inherent bilinear structure will be pointed out via transformation to a fixed-domain problem. $l(t)$ may occur as a *control variable* in various applications such as cranes or industrial roboters. A more complicated variant results from the problem

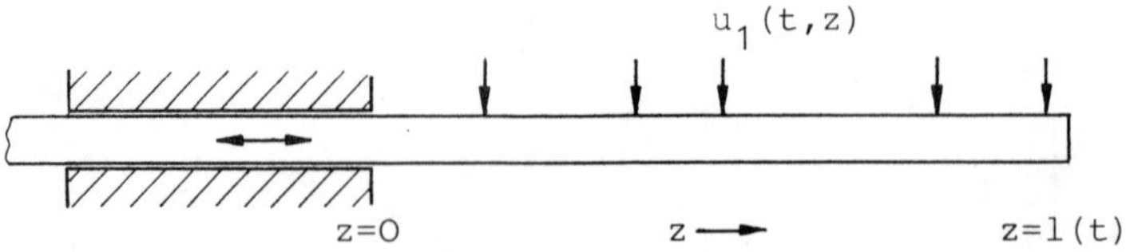

Fig.2.5: Vibrating beam with length $l(t)$ acting as a control variable

of elastic hauling pipes for deep sea mining [47], when the process of *letting down* the long pipe is considered.

Let $u_1(t,z)$ be the lumped or distributed load and $x(t,z)$ the transverse deflection. Then a proper state space representation for the undamped beam equation with state variables

$$x_1(t,z) = \partial^2 x/\partial z^2, \qquad x_2(t,z) = \partial x/\partial t \tag{2.26}$$

is given by [18], [24]:

$$\begin{aligned} \frac{\partial x_1}{\partial t} &= \frac{\partial^2 x_2}{\partial z^2}, \qquad 0 \leq z \leq l(t). \\ \frac{\partial x_2}{\partial t} &= -\frac{\partial^2 x_1}{\partial z^2} + u_1, \end{aligned} \tag{2.27}$$

with appropriate boundary conditions which may be omitted here.

The application of transformation

$$\zeta = z/l(t) \tag{2.28}$$

now provides the fixed-domain system

$$\begin{aligned} \frac{\partial x_1}{\partial t} &= \zeta \frac{\dot{l}}{l} \frac{\partial x_1}{\partial \zeta} + \frac{1}{l^2} \frac{\partial^2 x_2}{\partial \zeta^2}, \qquad 0 \leq \zeta \leq 1. \\ \frac{\partial x_2}{\partial t} &= \zeta \frac{\dot{l}}{l} \frac{\partial x_2}{\partial \zeta} - \frac{1}{l^2} \frac{\partial^2 x_1}{\partial \zeta^2} + u_1. \end{aligned} \tag{2.29}$$

This can again be interpreted as a bilinear distributed parameter system with linear input $u_1(t,z)$ and multiplicative control variables

$$u_2(t) = \dot{l}(t)/l(t), \qquad u_3(t) = 1/l^2(t). \tag{2.30}$$

u_2 and u_3 are of course interdependent.

2.3 State equations of bilinear distributed parameter systems

The material presented in sections 2.1 and 2.2 may suffice to illustrate the necessity of investigating the significant class of bilinear distributed parameter systems. The need of stating structural properties and efficient control methods is obvious. Therefore, it will be expedient to generalize the concrete examples by defining the state space representation of a wide class of bilinear distributed parameter systems as follows:

$$\frac{\partial \underline{x}(t,\underline{z})}{\partial t} = \underline{A}\,\underline{x}\,(t,\underline{z}) + \underline{u}_{\Omega}(t,\underline{z}), \qquad \underline{z}\in\Omega, \tag{2.31}$$

where

$$\underline{u}_{\Omega}(t,\underline{z}) = \sum_{i=1}^{p_1} u_i(t)\cdot\underline{B}_i\underline{x}(t,\underline{z}) + \underline{C}(\underline{z})\cdot\underline{u}^{(1)}(t); \tag{2.32}$$

boundary conditions:

$$\underline{R}\,\underline{x}\,(t,\underline{z}) = \underline{u}_{\Gamma}(t,\underline{z}), \qquad \underline{z}\in\Gamma, \tag{2.33}$$

where

$$\underline{u}_{\Gamma}(t,\underline{z}) = \sum_{i=p_1+1}^{p} u_i(t)\cdot\underline{B}_i\underline{x}(t,\underline{z}) + \underline{D}(\underline{z})\cdot\underline{u}^{(2)}(t); \tag{2.34}$$

initial condition:

$$\underline{x}(0,\underline{z}) = \underline{x}_o(\underline{z}), \qquad \underline{z}\in\Omega. \tag{2.35}$$

In many situations, especially in feedback control problems, an output equation has to be formulated:

$$\underline{y}(t) = \underline{E}\,\underline{x}\,(t,\underline{z}). \tag{2.36}$$

In the foregoing equations, $\underline{x}(t,\underline{z})$ is a n-dimensional vector representing the state of the system, $\underline{z}$ is a μ-dimensional spatial coordinate vector defined over a fixed spatial domain Ω, where Ω is a simple connected open region in μ-dimensional Euclidean space, Γ denotes its boundary, and t is time.

The p-dimensional control vector

$$\underline{u}(t) = [u_1(t), \ldots, u_{p_1}(t), u_{p_1+1}(t), \ldots, u_p(t)]^T =$$

$$= \begin{bmatrix} \underline{u}^{(1)}(t) \\ \\ \underline{u}^{(2)}(t) \end{bmatrix} \qquad (2.37)$$

is assumed to be lumped-type with regard to practical implementation, $\underline{u}^{(1)}$ acting in the interior of the system and $\underline{u}^{(2)}$ on its boundary. $\underline{A}$, $\underline{B}_i$, i=1, ..., p, and $\underline{R}$ are linear partial differential matrix operators with respect to the spatial coordinates $\underline{z}$. $\underline{C}(\underline{z})$ and $\underline{D}(\underline{z})$ are matrices whose elements are functions of $\underline{z}\in\Omega$ and $\underline{z}\in\Gamma$, respectively. $\underline{E}$ will, in general, be a matrix integral operator over the spatial variables $\underline{z}$. In case of linear pointwise measurement, which is of greatest practical importance, the matrix kernel of $\underline{E}$ contains spatial Dirac functions. $\underline{y}(t)$ is the q-dimensional output vector.

Eqs.(2.31) - (2.36) can readily be arranged to a fundamental block diagram outlined in Fig. 2.6.

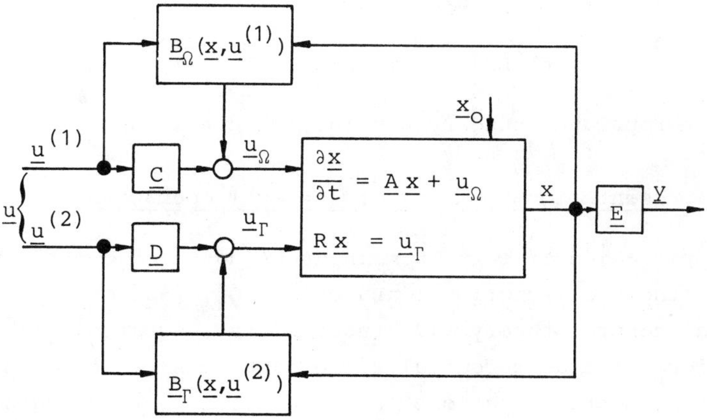

Fig.2.6: Structure of the state equations of a bilinear distributed parameter system

$$\underline{B}_\Omega(\underline{x},\underline{u}^{(1)}) = \sum_{i=1}^{p_1} u_i \cdot \underline{B}_i\underline{x}, \quad \underline{z}\in\Omega$$

$$\underline{B}_\Gamma(\underline{x},\underline{u}^{(2)}) = \sum_{i=p_1+1}^{p} u_i \cdot \underline{B}_i\underline{x}, \quad \underline{z}\in\Gamma$$

The definition (2.31) - (2.36) of a bilinear distributed parameter system may be further specialized by introducing additional hypotheses. Bilinear distributed systems will, as in the lumped parameter case [4], be defined to be *time varying* or *time invariant* according to wether or not $\underline{A}$, $\underline{B}_i$, $\underline{C}$, $\underline{D}$, $\underline{E}$, $\underline{R}$ depend on time t. Only time invariant systems will be regarded in this contribution.

A bilinear distributed parameter system will be said to be *homogeneous-in-the-state* if both $\underline{C} = \underline{0}$ and $\underline{D} = \underline{0}$, *homogeneous-in-the-input* if $\underline{A} = \underline{0}$, and *strictly bilinear* if $\underline{C} = \underline{D} = \underline{A} = \underline{0}$. This classification is of importance especially for the investigation of structural properties such as stability, reachability and controllability (chapter 3).

The main part of this contribution will not deal with the very general system description according to Eqs.(2.31) - (2.36). Much interest will be focussed on the scalar equation in one spatial coordinate z. Then, if in addition $p_1 = 1$, Eq.(2.31) along with Eq.(2.32) takes the simple form

$$\frac{\partial x}{\partial t} = Ax + u_1(t) \cdot Bx + C(z) \cdot u_1(t), \quad 0 \leqq z \leqq 1, \tag{2.38}$$

with appropriate boundary and initial conditions.

2.4 Representation by means of integral operators

The application of certain methods of advanced control theory, e.g. methods of functional analysis [40], [53] or Butkovskiy's optimal control theory [17] require state equations in integral form. The differential state equations can be transformed into integral equations in those cases where certain Green's function matrices can be constructed.

Let $\underline{G}(t-\tau,\underline{z},\underline{\zeta})$ be the Green's matrix [18] corresponding to the linear operator $(\underline{I} \cdot \partial/\partial t - \underline{A})$ along with boundary operator $\underline{R}$.

Then

$$\underline{x}(t,\underline{z}) = \int_{\tau=0}^{t} \int_{\Omega} \underline{G}(t-\tau,\underline{z},\underline{\zeta})\,\underline{u}_{\Omega}(\tau,\underline{\zeta})\,d\tau\, d\Omega_{\underline{\zeta}} +$$

$$+ \int_{\tau=0}^{t} \oint_{\Gamma} \underline{L}\{\underline{G}(t-\tau,\underline{z},\underline{\zeta})\} \underline{u}_{\Gamma}(\tau,\underline{\zeta}) d\tau d\Gamma_{\underline{\zeta}} +$$

$$+ \int_{\Omega} \underline{G}(t,\underline{z},\underline{\zeta}) \underline{x}_{o}(\underline{\zeta}) d\Omega_{\underline{\zeta}} . \tag{2.39}$$

Here, $\underline{L}$ is a linear operator with respect to $\underline{\zeta}$ whose form depends on $\underline{R}$. As $\underline{u}_{\Omega}$ and $\underline{u}_{\Gamma}$ have to be inserted according to Eqs. (2.32) and (2.34), respectively, Eq.(2.39) is in general an integro-differential equation. Integration by parts will help in some cases to obtain a proper integral equation. If all $\underline{B}_i$, $i=1, \ldots, p$, are matrices whose elements are *functions* of $\underline{z}$ (degenerated differential operators), integration by parts is of course dispensable.

Fig.2.7 illustrates the structure which is equivalent to that of Fig.2.6. Abbreviations $\underline{I}_{\Omega}$, $\underline{I}_{\Gamma}$, $\underline{I}_{o}$ have been used in Fig.2.7 for the linear integral operators in Eq.(2.39).

Two scalar examples may illustrate the nature of the integral state equations of bilinear systems.

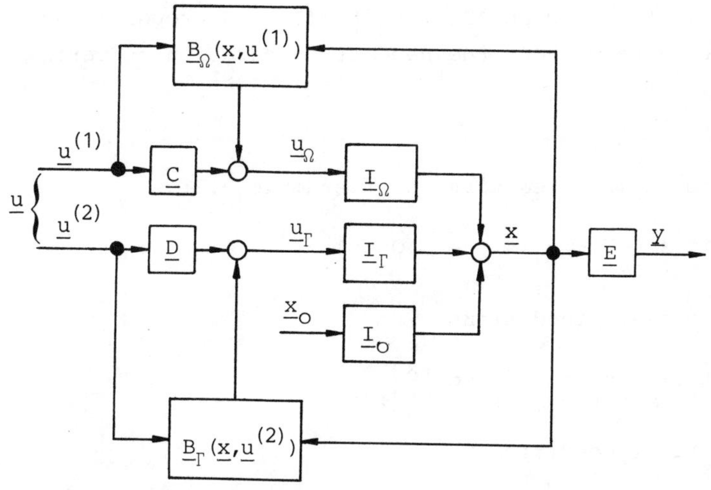

Fig.2.7: Structure of the integro-differential or integral state equations of a bilinear distributed system

Example 2.6:

Consider the boundary cooling process modelled by Eqs.(2.1o) - (2.13). Here the Green's function is given by [18]

$$g(t-\tau,z,\zeta) = \sum_{i=1}^{\infty} e^{\lambda_i(t-\tau)} \varphi_i(z)\varphi_i(\zeta), \tag{2.40}$$

where

$$\varphi_1(z) \equiv 1; \quad \varphi_i(z) = \sqrt{2}\cos(i-1)\pi z, \quad i = 2, 3, \ldots, \tag{2.41}$$

$$\lambda_i = -a(i-1)^2\pi^2, \quad i = 1, 2, \ldots, \tag{2.42}$$

and an integral equation for $x(t,z)$ is obtained as

$$\begin{aligned} x(t,z) = & \int_{\tau=0}^{t}\int_{\zeta=0}^{1} g(t-\tau,z,\zeta)u_1(\tau,\zeta)d\tau d\zeta - \\ & - a\int_{\tau=0}^{t} g(t-\tau,z,0)u_2(\tau)x(\tau,0)d\tau + \\ & + a\int_{\tau=0}^{t} g(t-\tau,z,1)u_3(\tau)d\tau + \int_{\zeta=0}^{1} g(t,z,\zeta)x_o(\zeta)d\zeta . \end{aligned} \tag{2.43}$$

Because of $B_\Omega = 0$ and $B_\Gamma = u_2(t)x(t,0)$ a proper integral equation results without the need of integration by parts.

Example 2.7:

Consider a modified version of example 2.2:

$$\frac{\partial x}{\partial t} = a\frac{\partial^2 x}{\partial z^2} - u_1(t)\frac{\partial x}{\partial z}, \quad 0 \leq z \leq 1, \tag{2.44}$$

with boundary conditions

$$x(t,0) = u_2(t), \quad \left[x + c\frac{\partial x}{\partial z}\right]_{z=1} = 0 \tag{2.45}$$

and initial condition

$$x(0,z) = x_o(z). \tag{2.46}$$

Again the Green's function $g(t-\tau,z,\zeta)$ corresponding to the linear operator $(\partial/\partial t - a\partial^2/\partial z^2)$ can be determined by known methods.

Now

$$x(t,z) = -\int_{\tau=0}^{t}\int_{\zeta=0}^{1} g(t-\tau,z,\zeta)u_1(\tau)\frac{\partial x}{\partial\zeta}\,d\tau d\zeta + $$
$$+ a\int_{\tau=0}^{t}\left.\frac{\partial g(t-\tau,z,\zeta)}{\partial\zeta}\right|_{\zeta=0}\cdot u_2(\tau)d\tau + \int_{\zeta=0}^{1} g(t,z,\zeta)x_o(\zeta)d\zeta. \tag{2.47}$$

The first integral in Eq.(2.47) contains the unfavourable term $\partial x/\partial\zeta$. Integration by parts yields

$$-\int_{\tau=0}^{t}\int_{\zeta=0}^{1} g(t-\tau,z,\zeta)u_1(\tau)\frac{\partial x}{\partial\zeta}\,d\tau d\zeta =$$
$$= \int_{\tau=0}^{t}\int_{\zeta=0}^{1}\frac{\partial g(t-\tau,z,\zeta)}{\partial\zeta}u_1(\tau)x(\tau,\zeta)d\tau d\zeta +$$
$$+ \int_{\tau=0}^{t} g(t-\tau,z,0)u_1(\tau)u_2(\tau)d\tau - \int_{\tau=0}^{t} g(t-\tau,z,1)u_1(\tau)x(\tau,1)d\tau, \tag{2.48}$$

which makes Eq.(2.47) a proper integral equation for $x(t,z)$. The kernel $\partial g/\partial\zeta$ is discontinuous at $\zeta = z$, due to the second order differential operator.

3 Remarks on structural properties

3.1 Equilibrium set and state equations in terms of perturbations

For the class of systems described by state equations (2.31) - (2.36), the determination of the set of equilibrium states $\underline{x}_s(\underline{z},\underline{u}_s)$ subject to some admissible set of fixed inputs $\underline{u}_s$ will be discussed. In a second step the *dynamic* state equations will be rewritten in terms of perturbations about a steady state. This is a preparation convenient for the application of certain methods such as Ljapunow's direct method.

The steady state equations take the form

$$\left(\underline{A} + \sum_{i=1}^{p_1} u_{is}\underline{B}_i\right)\underline{x}_s(\underline{z}) = -\underline{C}(\underline{z})\cdot\underline{u}_s^{(1)}, \quad \underline{z}\in\Omega, \tag{3.1}$$

with boundary conditions

$$\left(\underline{R} - \sum_{i=p_1+1}^{p} u_{is}\underline{B}_i\right)\underline{x}_s(\underline{z}) = \underline{D}(\underline{z})\cdot\underline{u}_s^{(2)}, \quad \underline{z}\in\Gamma. \tag{3.2}$$

Obviously, this is a *linear* boundary value problem, $\underline{u}_s$ playing the role of an input vector and, in addition, of a *parameter vector*.

Now let S denote the set of admissible inputs $\underline{u}$, and assume that for all $\underline{u}_s\in S$ the Green's matrix $\underline{G}_s(\underline{z},\underline{\zeta},\underline{u}_s)$ belonging to the operator

$$\left(\underline{A} + \sum_{i=1}^{p_1} u_{is}\underline{B}_i\right)$$

along with boundary operator

$$\left(\underline{R} - \sum_{i=p_1+1}^{p} u_{is}\underline{B}_i\right)$$

exists. Of course, $\underline{G}_s$ will depend nonlinearly on the components of $\underline{u}_s$.

Then the unique solution to the steady state equations can be expressed by

$$\underline{x}_s(\underline{z},\underline{u}_s) = -\int_\Omega \underline{G}_s(\underline{z},\underline{\zeta},\underline{u}_s)\underline{C}(\underline{\zeta})\,d\Omega_{\underline{\zeta}}\cdot\underline{u}_s^{(1)} + \\ + \oint_\Gamma \underline{L}\{\underline{G}_s(\underline{z},\underline{\zeta},\underline{u}_s)\}\underline{D}(\underline{\zeta})\,d\Gamma_{\underline{\zeta}}\cdot\underline{u}_s^{(2)}. \tag{3.3}$$

Again $\underline{L}$ is a linear partial differential operator with respect to $\underline{\zeta}$ whose form depends on the boundary operator.

The equilibrium set is the set of all functions $\underline{x}_s(\underline{z},\underline{u}_s)$ with $\underline{u}_s$ belonging to the admissible set S, according to Eq. (3.3).

If for some $\underline{\hat{u}}_s \in S$ the Green's matrix does not exist, no equilibrium state exists corresponding to $\underline{\hat{u}}_s$ unless $\underline{C}(\underline{z})\underline{\hat{u}}_s^{(1)}$ and $\underline{D}(\underline{z})\underline{\hat{u}}_s^{(2)}$ happen to lie in the range of the underlying operator, in which case the equilibrium set is augmented by an infinite number of steady states (example 3.2).

Obviously, $\underline{x}_s(\underline{z}) = \underline{0}$ is in any case contained in the equilibrium set of a homogeneous-in-the-state bilinear system.

Example 3.1:

A simple version of example 2.2 may be given by

$$\frac{\partial x}{\partial t} = - u_1(t)\cdot\frac{\partial x}{\partial z} - cx(t,z) + cu_2(t), \quad 0\leq z\leq 1, \tag{3.4}$$

with boundary condition

$$x(t,0) = u_3(t). \tag{3.5}$$

$c>0$ is a constant, $u_1(t)$ may be restricted by

$$(0\leq)m_1 \leq u_1(t) \leq M_1. \tag{3.6}$$

The corresponding equilibrium state problem is

$$u_{1s}x_s'(z) + cx_s(z) = cu_{2s},$$

$$x_s(0) = u_{3s},$$

and has as solution the equilibrium set:

$$x_s(z,\underline{u}_s) = [1-\exp(-\frac{c}{u_{1s}}z)]u_{2s} + \exp(-\frac{c}{u_{1s}}z)u_{3s}. \tag{3.7}$$

Example 3.2:

Consider example 2.3 with $u_1 \equiv 0$ and $(0\leq)m_2\leq u_2(t)\leq M_2$. Then if $m_2>0$ the Green's function of the steady state problem exists for all $u_{2s}\in[m_2,M_2]$, and the equilibrium set can readily be obtained as

$$x_s(z,\underline{u}_s) = u_{3s}z + \frac{u_{3s}}{u_{2s}}. \tag{3.8}$$

But if $m_2 = 0$ and $u_{2s} = 0$ (thermal insulation at $z=0$), the

Green's function does no longer exist, hence the system has no equilibrium state corresponding to $u_{2s} = 0$ unless $u_{3s} = 0$. In the latter case the above equilibrium set is augmented by the infinite number of steady states

$$x_s(z,\underline{0}) = c_1 \quad (c_1 \text{ arbitrary}). \tag{3.9}$$

Once the equilibrium state $\underline{x}_s(\underline{z})$ of a bilinear distributed system has been determined for a specified $\underline{u}_s$, it is sometimes convenient to rewrite the dynamic state equations in terms of perturbations.

Let

$$\underline{x}(t,\underline{z}) = \underline{x}_s(\underline{z}) + \Delta\underline{x}(t,\underline{z}), \tag{3.10}$$

$$\underline{u}(t) = \underline{u}_s + \Delta\underline{u}(t) \tag{3.11}$$

be the state and the input of the dynamical system, respectively, with (not necessarily small) perturbations $\Delta\underline{x}(t,\underline{z})$ and $\Delta\underline{u}(t)$. Then by means of simple insertion the state equations can be rewritten in terms of perturbations as

$$\frac{\partial\Delta\underline{x}}{\partial t} = \underline{\hat{A}}\Delta\underline{x} + \sum_{i=1}^{p_1}\Delta u_i\cdot\underline{B}_i\Delta\underline{x} + \underline{\hat{C}}(\underline{z})\cdot\Delta\underline{u}^{(1)}, \quad \underline{z}\in\Omega, \tag{3.12}$$

where

$$\underline{\hat{A}} = \underline{A} + \sum_{i=1}^{p_1} u_{is}\underline{B}_i, \tag{3.13}$$

$$\underline{\hat{C}}(\underline{z})\Delta\underline{u}^{(1)} = \underline{C}(\underline{z})\Delta\underline{u}^{(1)} + \sum_{i=1}^{p_1}\Delta u_i\underline{B}_i\underline{x}_s(\underline{z}); \tag{3.14}$$

boundary conditions:

$$\underline{\hat{R}}\Delta\underline{x} = \sum_{i=p_1+1}^{p}\Delta u_i\cdot\underline{B}_i\Delta\underline{x} + \underline{\hat{D}}(\underline{z})\cdot\Delta\underline{u}^{(2)}, \quad \underline{z}\in\Gamma, \tag{3.15}$$

where

$$\underline{\hat{R}} = \underline{R} - \sum_{i=p_1+1}^{p} u_{is}\underline{B}_i, \tag{3.16}$$

$$\underline{\hat{D}}(\underline{z})\Delta\underline{u}^{(2)} = \underline{D}(\underline{z})\Delta\underline{u}^{(2)} + \sum_{i=p_1+1}^{p}\Delta\underline{u}_i\underline{B}_i\underline{x}_s(\underline{z}); \tag{3.17}$$

initial condition:

$$\Delta \underline{x}(0,\underline{z}) = \Delta \underline{x}_o(\underline{z}), \qquad \underline{z} \in \Omega; \tag{3.18}$$

output equation ($\underline{E}$ linear):

$$\Delta \underline{y}(t) = \underline{E} \Delta \underline{x}(t,\underline{z}). \tag{3.19}$$

It can be observed that transition to perturbations does not change the bilinear nature of the system. Moreover, bilinear terms reproduce, whereas linear terms in the state equations and in the boundary conditions are modified ($\hat{}$); they are essentially effected by the pre-determined steady state.

In order to keep notations simple, Δ and ($\hat{}$) will be omitted in the following whenever confusion can be excluded. Of course, the steady state of the perturbed equations is zero.

Example 3.3:

Example 3.1, rewritten in terms of perturbations, is

$$\frac{\partial x}{\partial t} = - cx(t,z) - u_{1s} \frac{\partial x}{\partial z} - u_1(t) \frac{\partial x}{\partial z} - x_s'(z) u_1(t) + cu_2(t), \qquad 0 \leq z \leq 1; \tag{3.20}$$

boundary condition:

$$x(t,0) = u_3(t); \tag{3.21}$$

initial condition:

$$x(0,z) = x_o(z); \tag{3.22}$$

output equation (possibly):

$$y(t) = x(t,1). \tag{3.23}$$

Example 3.4:

Consider example 2.3 with $u_1 \equiv 0$ and $u_{2s} \neq 0$ which makes the steady state solution unique according to Eq.(3.8). Then Eqs. (2.10)-(2.13), rewritten in terms of perturbations, assume the form

$$\frac{\partial x}{\partial t} = a \frac{\partial^2 x}{\partial z^2}, \quad 0 \leq z \leq 1; \tag{3.24}$$

$$\left[\frac{\partial x}{\partial z} - u_{2s} x(t,z)\right]_{z=0} = u_2(t)x(t,0) + x_s(0)u_2(t), \tag{3.25}$$

$$\left.\frac{\partial x}{\partial z}\right|_{z=1} = u_3(t), \tag{3.26}$$

$$x(0,z) = x_0(z); \tag{3.27}$$

output equation (possibly):

$$y(t) = x(t,0). \tag{3.28}$$

3.2 Stability analysis

Stability, as a fundamental structural property, will briefly be discussed in this section for a class of open-loop bilinear distributed parameter processes. The *synthesis* of stable closed-loop control systems will be the object of chapter 5.

Consider the following homogeneous-in-the-state bilinear system, with vector-valued state $\underline{x}(t,\underline{z})$ and control vector $\underline{u}(t)$, taking values from some admissible set S.

$$\frac{\partial \underline{x}}{\partial t} = \left(\underline{A} + \sum_{i=1}^{p_1} u_i \cdot \underline{B}_i\right)\underline{x}, \quad \underline{z} \in \Omega, \tag{3.29}$$

$$\underline{R}\,\underline{x} = \sum_{i=p_1+1}^{p} u_i \cdot \underline{B}_i \underline{x}, \qquad \underline{z} \in \Gamma, \tag{3.30}$$

with initial state

$$\underline{x}(0,\underline{z}) = \underline{x}_0(\underline{z}). \tag{3.31}$$

With regard to section 3.1, the above system will not necessarily have a unique steady state for fixed input $\underline{u}_s \in S$, but

$$\underline{x}_s(\underline{z}) \equiv \underline{0} \tag{3.32}$$

will in any case be in the equilibrium set.

Therefore, the question arises if for arbitrary $\underline{u}(t) \in S$ the system's motion, starting from $\underline{x}_0(\underline{z})$, will tend to the zero equilibrium state as time t increases. Hence, the notion of *Ljapunow-stability* will be used, and Ljapunow's direct method, as extended to distributed parameter systems by Zubow [61], Massera [49] and Persidskii [52], is an appropriate tool to obtain sufficient stability conditions for a specified system.

It should be noticed that *BIBO-stability* with respect to *step input* of non-homogeneous bilinear systems can be treated by the same methods. If the step input vector is $\underline{u}(t) = \underline{u}_s\sigma(t)$, where $\sigma(t)$ denotes the unit step function, and if an equilibrium state $\underline{x}_s(\underline{z})$ exists corresponding to $\underline{u}_s$, then BIBO-stability of the step input response can be reduced to a problem of Ljapunow-stability by rewriting the state equations in terms of perturbations about $\underline{x}_s(\underline{z})$.

A tentative Ljapunow-functional $V(\underline{x})$ may be formulated as

$$V(\underline{x}) = \int_\Omega \underline{x}^T(t,\underline{z})\underline{P}\,\underline{x}\,(t,\underline{z})\,d\Omega, \tag{3.33}$$

with positive definite matrix $\underline{P}$. $V(\underline{x})$ is continuous with respect to the metric

$$\rho(\underline{x},\underline{0}) = \|\underline{x}\| = \sqrt{\int_\Omega \underline{x}^T\underline{x}\,d\Omega}. \tag{3.34}$$

Now by means of Eq.(3.29)

$$\dot{V} = \int_\Omega \left[\frac{\partial \underline{x}^T}{\partial t}\underline{P}\,\underline{x} + \underline{x}^T\underline{P}\frac{\partial \underline{x}}{\partial t}\right]d\Omega =$$

$$= \int_\Omega \left\{\left[(\underline{A} + \sum_{i=1}^{p_1} u_i\underline{B}_i)\underline{x}\right]^T\underline{P}\,\underline{x} + \underline{x}^T\underline{P}(\underline{A} + \sum_{i=1}^{p_1} u_i\underline{B}_i)\underline{x}\right\}d\Omega. \tag{3.35}$$

In concrete examples, integration by parts will allow the boundary conditions, Eq.(3.30), to be inserted and then to examine if $\dot{V} \leqq 0$ for all $\underline{u}(t) \in S$.

$\underline{x}_s(\underline{z}) = \underline{0}$ is assured to be *stable* with respect to metric (3.34) if for any $\underline{u}(t) \in S$ and for all $\underline{x}_o(\underline{z})$ satisfying $\|\underline{x}_o\| \leqq r$ the functional $\dot{V}$ is negative *semidefinite*, where r is a sufficiently small positive number.

If, in addition, $\dot{V} \not\equiv 0$ along any perturbed motion or even $\dot{V}$ negative *definite*, then $\underline{x}_s(\underline{z}) = \underline{0}$ is assured to be *asymptotically stable* with respect to metric (3.34), and $\|\underline{x}\| \leqq r$ belongs to the domain of asymptotic stability.

The procedure will be illustrated using two simple practical examples.

Example 3.5:

Consider the very trivial example 2.1 with additional assumption that $u_2 \equiv 0$. Here $A = 0$ and $B_1 = -\partial/\partial z$. Let the velocity $u_1(t)$ of the conveyor-belt be bounded by

$$S = \{u_1(t) \mid (0 \leqq) m_1 \leqq u_1(t) \leqq M_1\}. \tag{3.36}$$

Now by means of the Ljapunow-functional

$$V = \frac{1}{2}\int_0^1 x^2(t,z)\,dz \tag{3.37}$$

we obtain

$$\dot{V} = \int_0^1 x \frac{\partial x}{\partial t}\,dz = -u_1(t)\int_0^1 x \frac{\partial x}{\partial z}\,dz = -\frac{1}{2}u_1(t)x^2(t,1) \tag{3.38}$$

which is, of course, negative semidefinite for any $u_1(t) \in S$, thus making a stable system in the sense of Ljapunow, even if $m_1 = 0$ and $u_1(t) \equiv 0$.

If $m_1 > 0$, $\dot{V}$ is still negative *semi*definite, but $\dot{V} \equiv 0$ now requires $x(t,1) \equiv 0$ and therefore $x_0(z) \equiv 0$. This makes $\dot{V} \not\equiv 0$ along any perturbed motion, and therefore the zero state is asymptotically stable for arbitrary $x_0(z)$. This result can, of course, be obtained immediately for physical reasons.

Example 3.6:

Bilinear boundary control example 2.3 is considered with additional assumptions that $u_1 \equiv 0$ and $u_3 \equiv 0$, and $u_2(t)$ being bounded by

$$S = \{u_2(t) \mid (0 \leqq) m_2 \leqq u_2(t) \leqq M_2\}. \tag{3.39}$$

By choosing again

$$V = \frac{1}{2}\int_0^1 x^2(t,z)\,dz,$$

now

$$\dot{V} = a\int_0^1 x\,\frac{\partial^2 x}{\partial z^2}\,dz = a\left[x\cdot\frac{\partial x}{\partial z}\right]_0^1 - a\int_0^1\left(\frac{\partial x}{\partial z}\right)^2 dz =$$

$$= -\,au_2(t)x^2(t,0) - a\int_0^1\left(\frac{\partial x}{\partial z}\right)^2 dz, \tag{3.40}$$

which is negative semidefinite for any $u_2(t)\in S$, thus making a stable system in the sense of Ljapunow, even if $m_2 = 0$ and $u_2(t) \equiv 0$ (thermal insulation).

If $m_2 > 0$, $\dot{V} \equiv 0$ requires both $x(t,0) \equiv 0$ and $\partial x/\partial z \equiv 0$, hence $x(t,z) \equiv 0$, which makes $\dot{V}$ negative definite and therefore the zero state asymptotically stable for arbitrary $x_o(z)$.

These examples may seem to be rather trivial, but similar techniques will serve for the design of effective feedback control laws in chapter 5.

3.3 Reachability and controllability

The notions of reachability and controllability are intrinsic structural properties which a given dynamical system may have or not. Moreover, these properties are closely related to the question of well-posedness of certain optimal control problems and to the profit of optimal control. It is the merit of Mohler [11], [12] to have shown in the lumped parameter case that bilinear systems, due to their adaptive nature, are not only "more controllable" than linear systems, but that also the performance of optimal control can considerably be improved by introducing multiplicative control modes. It can therefore be expected that bilinear distributed systems exhibit similar advantages.

As far as lumped parameter systems are concerned, several reachability statements and controllability conditions are available, especially in the strictly bilinear and in the bilinear-in-the-state case. The answers can be formulated in terms of finite-dimensional matrices and therefore have an intrinsic character. It seems that in the distributed case statements of that generality can hardly be obtained. Therefore, only an exemplary discussion will be given below after some preliminary definitions.

3.3.1 Definitions

The bilinear system, described by Eqs.(2.31) - (2.35) is considered, with bounded $\underline{u}(t) \in S$.

A state $\hat{\underline{x}}(\underline{z})$ is said to be *reachable* from $\underline{x}_o(\underline{z})$ if there exists a control $\underline{u}(t) \in S$ which steers the system from $\underline{x}_o(\underline{z})$ to $\hat{\underline{x}}(\underline{z})$ in a finite time T. The set of states reachable from $\underline{x}_o$ in time T will be denoted as $X_T(\underline{x}_o)$. Then the *reachable set* from $\underline{x}_o$, $\hat{X}(\underline{x}_o)$, is defined as

$$\hat{X}(\underline{x}_o) = \bigcup_{0 \leq T < \infty} X_T(\underline{x}_o) . \tag{3.41}$$

A system is *controllable* at $\underline{x}_o(\underline{z})$ if $\hat{X}(\underline{x}_o)$ spans the total state space. It is *completely controllable* if it is controllable at any $\underline{x}_o(\underline{z})$.

A system is *locally controllable* at $\underline{x}_o(\underline{z})$ if $\hat{X}(\underline{x}_o)$ contains an ε-neighbourhood of $\underline{x}_o(\underline{z})$.

Kalman's [44] fundamental theorems on controllability of linear lumped parameter systems can be extended to a class of linear distributed systems [18]. It is significant to note, however, that complete controllability usually requires infinite control. In fact, similar to the lumped parameter case [12] it is easy to prove the following theorem:

The linear distributed system (2.31) - (2.35), where $\underline{B}_i = \underline{0}$, $i = 1, \ldots, p$, is *not* completely controllable with *bounded* $\underline{u}$, if the corresponding autonomous system is asymptotically stable.

To show this, first consider the autonomous linear system along with Ljapunow-functional (3.33) which will assure $\dot{V} = q_1(\underline{x})$ to be negative definite, where q_1 is a quadratic functional in $\underline{x}$. For the same system with added linear control term, we have

$$\dot{V} = q_1(\underline{x}) + q_2(\underline{x},\underline{u}) , \tag{3.42}$$

where q_2 is a functional linear in $\underline{x}$ and linear in $\underline{u}$. Now, $\dot{V}$ is negative for every bounded $\underline{u}$, with $\|\underline{x}\|$ and therefore $V(\underline{x}) = V^*$ sufficiently large. Hence, the system's motion cannot leave the domain bounded by $V(\underline{x}) = V^*$, and therefore the asymptotically stable linear system is not completely controllable with bounded control.

This fundamental relationship between stability and controllability has been exploited by Mohler to provide sufficient conditions for complete controllability of bilinear lumped systems. These conditions essentially require that all eigenvalues of the *constant* parameter system matrix of Eq.(1.2),

$$\underline{A} + \sum_{i=1}^{p} u_i \underline{B}_i,$$

can be shifted across the imaginary axis of the complex plane, as $\underline{u}$ ranges over the admissible set S.

Conditions like this one will hardly be transferable to the distributed parameter case. Butkovskiy [17], [39] has employed the method of moments to derive reachability conditions for linear distributed parameter systems with bounded control. This technique seems to be an appropriate tool also in the bilinear case and will be treated for a special class of systems in the following section.

3.3.2 Reachability for a class of bilinear distributed systems

Consider the class of bilinear systems describable by an integral equation of the type

$$x(t,z) = \int_{\tau=o}^{t} [g_1(t-\tau,z)u_1(\tau) + g_2(t-\tau,z)x(\tau,o)u_2(\tau)]d\tau + \\ + \int_{\zeta=o}^{1} g(t,z,\zeta)x_o(\zeta)d\zeta, \qquad z\in[o,1], \qquad (3.43)$$

with independent linear control u_1 and multiplicative boundary control u_2. Example 2.6, Eq.(2.43), with $u_3 \equiv o$ is of this type.

Let the set of admissible controls be given by

$$S = \{\underline{u}(t) | m_i \leq u_i(t) \leq M_i;\ i=1,2\}. \tag{3.44}$$

Then the fundamental question arises if a specified state $\hat{x}(z)$ is reachable from a given initial state $x_o(z)$. It can be concluded from Eq.(3.43) that for $\hat{x}(z)$ to be reachable from $x_o(z)$ in time T, a solution $\underline{u}(t) \in S$ of the following integral equation must exist:

$$\hat{x}(z) - \int_{\zeta=o}^{1} g(t,z,\zeta) x_o(\zeta) d\zeta =$$

$$= \int_{\tau=o}^{T} [g_1(T-\tau,z) u_1(\tau) + g_2(T-\tau,z) x(\tau,0) u_2(\tau)] d\tau. \tag{3.45}$$

The question of existence can be given an answer of *sufficient* character by means of the method of moments.

Therefore, let

$$\{\varphi_k(z)\},\ z \in [0,1],\quad k = 1, 2, \ldots, \tag{3.46}$$

be an arbitrary complete set of functions and expand the functions

$$\hat{x}(z) - \int_{\zeta=o}^{1} g(T,z,\zeta) x_o(\zeta) d\zeta = \sum_{k=1}^{\infty} \alpha_k \varphi_k(z), \tag{3.47}$$

$$g_i(T-\tau,z) = \sum_{k=1}^{\infty} g_{ik}(T-\tau) \varphi_k(z),\ i=1,\ 2. \tag{3.48}$$

Inserting these expansions into Eq.(3.45) and comparing coefficients yields

$$\alpha_k = \int_{\tau=o}^{T} [g_{1k}(T-\tau) u_1(\tau) + g_{2k}(T-\tau) x(\tau,0) u_2(\tau)] d\tau,$$

$$k = 1,\ 2,\ \ldots \tag{3.49}$$

In the linear case ($u_2 \equiv 0$) this is known as an infinite-dimensional problem of moments [39].

Problem (3.49) can be made to be *formally* linear by means of the substitution

$$v(t) := x(t,0)u_2(t), \tag{3.50}$$

and by assuming that

$$m \leq v(t) \leq M \tag{3.51}$$

with yet unspecified bounds m and M.

Then validity of the following statement is obvious:

For $\hat{x}(z)$ to be reachable from $x_o(z)$ in time T it is sufficient that two numbers m and M can be found such that

(a) a solution to the following problem of moments exists:

$$\alpha_k = \int_{\tau=0}^{T} [g_{1k}(T-\tau)u_1(\tau) + g_{2k}(T-\tau)v(\tau)]d\tau, \tag{3.52}$$

$k = 1,2,\ldots,\ m_1 \leq u_1(t) \leq M_1,\ m \leq v(t) \leq M,$

and in addition, with regard to Eq.(3.50),

(b) the inequality

$$m \leq x(t,0)u_2(t) \leq M \tag{3.53}$$

holds for some $u_2(t)$, $m_2 \leq u_2(t) \leq M_2$, where

$$x(t,0) = \int_{\tau=0}^{t} [g_1(t-\tau,0)u_1(\tau)+g_2(t-\tau,0)v(\tau)]d\tau + \int_{\zeta=0}^{1} g(t,0,\zeta)x_o(\zeta)d\zeta \tag{3.54}$$

is available from subproblem (a).

Remark 1: The examination of the stated sufficient reachability condition in concrete situations generally will require an iterative procedure to adjust the numbers m and M.

Remark 2: If the given interval $[m_2,M_2]$ contains the origin, then the interval $[m,M]$ will contain the origin, too. In this case, it is sufficient for $\hat{x}(z)$ to be reachable from $x_o(z)$ in time T, if, by setting $u_2(t) \equiv 0$, the following simplified problem of moments has a solution:

$$\alpha_k = \int_{\tau=0}^{T} g_{1k}(T-\tau)u_1(\tau)d\tau, \quad k=1,\ 2,\ \ldots;\ m_1 \leq u_1(t) \leq M_1.$$

Moreover, additional application of multiplicative control u_2 may significantly enlarge the reachable set and thus make the bilinear system more controllable than the linear one.

Remark 3: If only multiplicative control u_2 is applied, $u_1(t) \equiv 0$, then the system is homogeneous-in-the-state. If then, additionally, the zero steady state is asymptotically stable, the reachable set will be very restricted (examples 3.5 and 3.6). Especially, if $\hat{x}(z)$ happens to be reachable from $x_o(z)$ in time T, it will lose this property for enlarged T.

4. Feedback control structures

In the foregoing sections bilinear distributed processes and some of their structural properties have been described excluding any feedback control. This chapter will be concerned with some feedback control structures of bilinear distributed systems. As in the lumped parameter case [4], linear state or output feedback of a bilinear system will make a *quadratic-in-the-state* dynamical system. In this context special interest will be focussed on *lumped* dynamic feedback devices with regard to practical implementation (section 4.1). Hence controlled processes with mixed lumped and distributed nature will be considered.

Becker [1], [2] has pointed out in the lumped parameter case that bilinear systems arise also in piecewise linear control of linear systems. Here the *parameters* of the controller are interpreted as multiplicative input variables. Becker has shown that the performance of such a variable structure control system is significantly better than that of fixed parameter control. Bellman's functional equation and Ljapunow's direct method have been applied in [2] for the near-optimal design.

The use of variable structure control in linear distributed parameter systems can be expected to be of similar efficiency as in the lumped parameter case. Therefore, a description of this appealing class of systems will be given in section 4.2.

No controller design will be carried out in chapter 4. This will be the object of chapters 5 and 6.

4.1 Linear lumped control of bilinear distributed systems

4.1.1 State equations of the mixed over-all system

Consider the bilinear distributed plant outlined in Fig. 2.6 to be controlled via linear lumped parameter subsystem (including sensor, controller and actuator). The structure of the resulting system is sketched in Fig. 4.1, which will in general be a multivariable system. It seems to be worth-while to investigate control systems of this type, as multiplicative control is not only easy to implement in many situations but proves to be most efficient with regard to the performance of the total system.

By assuming $p_1 = 1$ in Eq.(2.32) and $p = 2$ in Eq.(2.34) for simplifying notations, the state equations of the lumped subsystem in Fig. 4.1 will take the form

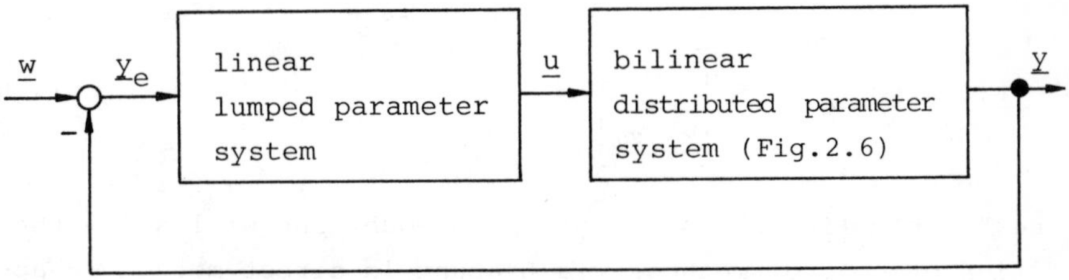

Fig.4.1: Structure of the multivariable control loop; $\underline{w}$ reference input, $\underline{y}_e$ error

$$\dot{\underline{\xi}}(t) = \tilde{\underline{A}}\,\underline{\xi}(t) + \tilde{\underline{B}}\,\underline{y}_e(t), \tag{4.1}$$

$$u_i(t) = \tilde{\underline{c}}_i^T \underline{\xi}(t), \; i = 1, 2, \tag{4.2}$$

where $\underline{\xi}$ is an m-dimensional state vector and $\tilde{\underline{A}}$, $\tilde{\underline{B}}$, $\tilde{\underline{c}}_i^T$ are constant matrices of suitable dimensions.

Now by inserting Eqs.(4.1) and (4.2) into Eqs. (2.31) - (2.36), the state equations of the closed-loop system are readily obtained as

$$\dot{\underline{\xi}} = \tilde{\underline{A}}\,\underline{\xi} + \tilde{\underline{B}}\,\underline{w} - \tilde{\underline{B}}\,\underline{E}\,\underline{x}, \tag{4.3}$$

$$\frac{\partial \underline{x}}{\partial t} = \underline{A}\,\underline{x} + \underline{\xi}^T\tilde{\underline{c}}_1\underline{B}_1\underline{x} + \underline{C}(\underline{z})\tilde{\underline{c}}_1^T\underline{\xi}, \; \underline{z}\in\Omega; \tag{4.4}$$

boundary conditions:

$$\underline{R}\,\underline{x} = \underline{\xi}^T\tilde{\underline{c}}_2\underline{B}_2\underline{x} + \underline{d}(\underline{z})\tilde{\underline{c}}_2^T\underline{\xi}, \; \underline{z}\in\Gamma; \tag{4.5}$$

initial conditions:

$$\underline{\xi}(0) = \underline{\xi}_o; \; \underline{x}(0,\underline{z}) = \underline{x}_o(\underline{z}). \tag{4.6}$$

The terms $\underline{\xi}^T\tilde{\underline{c}}_i\underline{B}_i\underline{x}$, i = 1, 2, indicate the *quadratic-in-the-state* nature of the obtained equations. By introducing the total state vector

$$\underline{\eta}(t,\underline{z}) = \begin{bmatrix} \underline{\xi}(t) \\ \underline{x}(t,\underline{z}) \end{bmatrix}, \tag{4.7}$$

Eqs. (4.3), (4.4) and (4.6) can be summarized as

$$\frac{\partial \underline{\eta}}{\partial t} = \begin{bmatrix} \tilde{\underline{A}} & -\tilde{\underline{B}}\,\underline{E} \\ \underline{C}(\underline{z})\tilde{\underline{c}}_1^T & \underline{A}+\underline{\xi}^T\tilde{\underline{c}}_1\underline{B}_1 \end{bmatrix} \underline{\eta} + \begin{bmatrix} \tilde{\underline{B}} \\ \underline{0} \end{bmatrix} \underline{w}; \tag{4.8}$$

$$\underline{\eta}(0,\underline{z}) = \underline{\eta}_o(\underline{z}) = \begin{bmatrix} \underline{\xi}_o \\ \underline{x}_o(\underline{z}) \end{bmatrix}. \tag{4.9}$$

The "nearly linear" nature of these equations will allow the design of control methods via Ljapunow's direct method (chapter 5).

4.1.2 Accuracy of equilibrium reference input response

In many situations the perturbed motions of the controlled system about a pre-assigned steady state are of interest. The determination of this steady state will be treated in advance apart from dynamical considerations.

Let $\underline{w}$ be a fixed reference input vector. Then the question of the relation between $\underline{w}$ and the steady state output $\underline{y}_s$ arises. It may be assumed that

(a) $p_1 = 1$, $p = 2$;

(b) $q = p$; (see notation)

(c) $\underline{\tilde{A}}$ be non-singular;

(d) $$\underline{K} = \begin{bmatrix} \underline{k}_1^T \\ \underline{k}_2^T \end{bmatrix} = - \begin{bmatrix} \underline{\tilde{c}}_1^T \\ \underline{\tilde{c}}_2^T \end{bmatrix} \underline{\tilde{A}}^{-1}\underline{\tilde{B}} \text{ be non-singular;}$$

(e) an equation of the type of Eq.(3.3) be valid for the equilibrium set of the distributed system.

Assumption (a) has been made just for simplifying notations.

The equilibrium output of the lumped part will then be described by

$$\underline{u}_s = \underline{K} \cdot (\underline{w} - \underline{y}_s), \tag{4.10}$$

and the equilibrium output of the distributed part, in view of Eqs.(3.3) and (2.36), will be described by a nonlinear vector-valued function

$$\underline{y}_s = \underline{f}(\underline{u}_s) \tag{4.11}$$

with general property

$$\underline{f}(\underline{0}) = \underline{0}. \tag{4.12}$$

Now by combining Eqs.(4.10) and (4.11) the closed loop equation is

$$\underline{y}_s = \underline{f}[\underline{K}\cdot(\underline{w}-\underline{y}_s)] \tag{4.13}$$

which, except for $\underline{w} = \underline{0}$, will never have solution $\underline{y}_s = \underline{w}$, due to Eq.(4.12). Indeed, by the assumptions made, any integral behaviour of the open-loop system has been excluded.

But, by means of a simple modification, equilibrium output accuracy $\underline{y}_s = \underline{w}$ can be achieved. Therefore, the input $\underline{w}$ will be replaced by a nonlinear vector-valued function $\underline{g}(\underline{w})$ which can be determined from the requirement that $\underline{y}_s = \underline{w}$:

$$\underline{w} = \underline{f}[\underline{K}\,\underline{g}\,(\underline{w}) - \underline{K}\,\underline{w}\,]. \tag{4.14}$$

This is an implicit equation for $\underline{g}(\underline{w})$ which, if Eq.(4.11) can be written as

$$\underline{u}_s = \underline{f}^{-1}(\underline{y}_s), \tag{4.15}$$

assumes the explicit form

$$\underline{g}(\underline{w}) = \underline{w} + \underline{K}^{-1}\underline{f}^{-1}(\underline{w}). \tag{4.16}$$

Example 4.1:

Consider example 3.1 with scalar multiplicative control variable $u_1(t)$, $0 \leq m_1 \leq u_1(t) \leq M_1$, and fixed inputs $u_2 = u_{2s}$, $u_3 = u_{3s} \neq u_{2s}$. Then if the scalar output is specified as

$$y(t) = Ex(t,z) = \int_0^1 \delta(z,1)x(t,z)dz = x(t,1), \tag{4.17}$$

the equilibrium output is obtained from Eq.(3.7) as

$$y_s = u_{2s} + (u_{3s}-u_{2s})\exp(-c/u_{1s}) = f(u_{1s}). \tag{4.18}$$

This equation can be solved, according to Eq.(4.15), as

$$u_{1s} = f^{-1}(y_s) = \frac{c}{\ln(u_{3s}-u_{2s})-\ln(y_s-u_{2s})}. \tag{4.19}$$

Now if K>0 is the scalar gain of the linear lumped controller, Eq.(4.16) takes the concrete form

$$g(w) = w + \frac{c/K}{\ln(u_{3s}-u_{2s})-\ln(w-u_{2s})}. \tag{4.20}$$

Fig. 4.2 illustrates this nonlinear "filter function" for the cooling process with $u_{2s} = 0$, $u_{3s} = 1$.

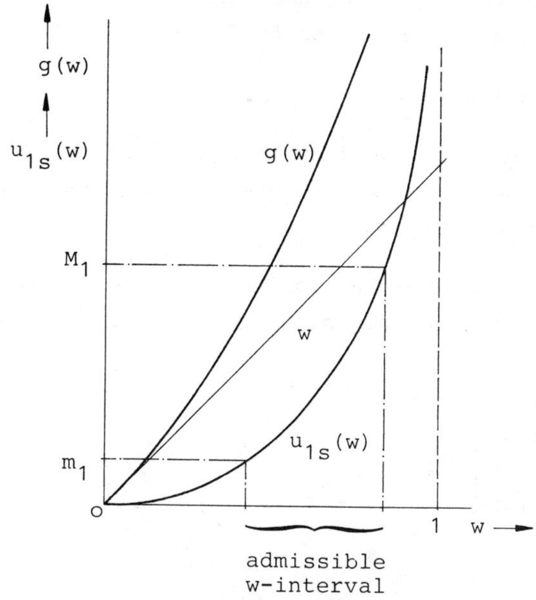

Fig.4.2: Nonlinear filter function g(w) in example 4.1

Of course, only $0 \leq w < 1$ makes a sense for physical reasons. Bounds on u_1 will even reduce the admissible w-interval as outlined in Fig. 4.2.

Example 4.2:

Let $u_2(t)$ act as a scalar multiplicative control variable in example 2.3. Moreover, let $u_1 \equiv 0$ and $u_3 = u_{3s} \neq 0$. Then if $0 < m_2 \leq u_2(t) \leq M_2$, the equilibrium set is given by Eq.(3.8).

Let the surface temperature $y(t) = x(t,0)$ be the scalar output of the system. Then the equilibrium output is

$$y_s = x_s(0) = u_{3s}/u_{2s} = f(u_{2s}), \tag{4.21}$$

hence

$$u_{2s} = f^{-1}(y_s) = u_{3s}/y_s \tag{4.22}$$

according to Eq.(4.15).

Now if K is the scalar gain of the linear lumped subsystem, Eq.(4.16) assumes the concrete form

$$g(w) = w + \frac{u_{3s}}{Kw} \quad . \tag{4.23}$$

It must be stressed that by intuition K will be made *negative* in this example such as to intensify the cooling effect if $y>w$ and to weaken it if $y<w$.

Fig. 4.3 illustrates the nonlinear function g(w) according to Eq.(4.23), and the admissible w-interval in view of bounded u_2.

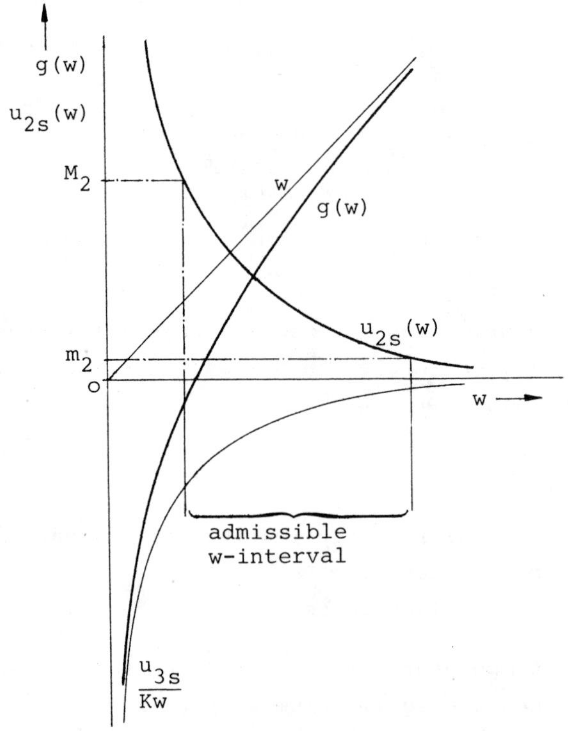

Fig.4.3: Nonlinear filter function g(w) in example 4.2

Of course, the question of stability of the closed-loop system has not been touched by the considerations of this section. This will be the object of chapter 5.

4.2 Variable structure control of linear distributed systems

Consider the multivariable control system sketched in Fig. 4.4. Different from Fig. 4.1, the distributed parameter plant is now assumed to be linear as well. Similar to the considerations by Becker, the over-all system assumes a bilinear nature via nonconstant matrix-type feedback gain. The system can be said to be parameter-controlled by time dependent gain matrix $\underline{K}(t)$. As in most contributions in the lumped parameter case [2], [8], [9], $\underline{K}(t)$ is supposed to be piecewise constant thus making the over-all system piecewise linear. Some authors have made successful use of linear systems theory for the design of near-optimal switching strategies for $\underline{K}(t)$. On the other hand, by interpreting the over-all system as a *bilinear* system, a new tool of treating variable structure control systems opens [1].

Two mathematical models of the system outlined in Fig.4.4 will be given below as a preparation of optimal variable structure control of these mixed-type systems to be considered in chapter 6. The reference input is assumed to be equal to zero for simplicity, hence perturbations about a pre-determined steady state being considered.

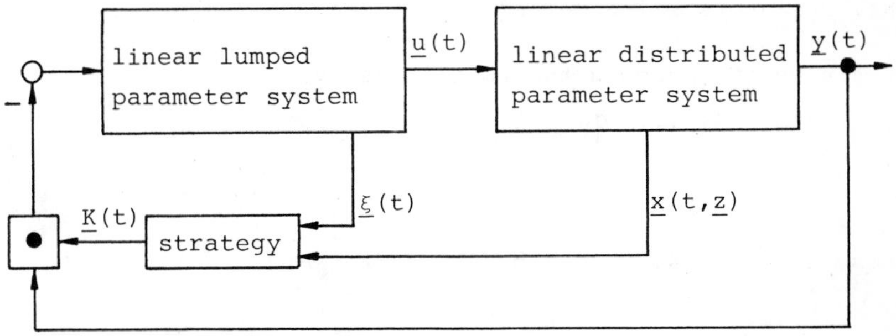

Fig.4.4: Variable structure control system

Lumped and possibly distributed state observers [46],[60] have to be taken into consideration depending on whether or not the total state vectors $\underline{\xi}(t)$ and $\underline{x}(t,\underline{z})$ of the lumped and distributed system, respectively, are needed for the switching strategy.

4.2.1 State equations of the mixed-type variable structure control system.

Linear lumped subsystem (including output feedback):

$$\dot{\underline{\xi}}(t) = \underline{\tilde{A}}\,\underline{\xi}(t) + \underline{\tilde{B}} \cdot \left[- \underline{K}(t)\underline{y}(t)\right] , \tag{4.24}$$

$$\underline{u}(t) = \begin{bmatrix} \underline{u}^{(1)}(t) \\ \underline{u}^{(2)}(t) \end{bmatrix} = \begin{bmatrix} \underline{\tilde{C}}^{(1)} \\ \underline{\tilde{C}}^{(2)} \end{bmatrix} \underline{\xi}(t) = \underline{\tilde{C}}\,\underline{\xi}(t), \tag{4.25}$$

$$\underline{\xi}(0) = \underline{\xi}_o. \tag{4.26}$$

Linear distributed subsystem, resulting as a special case from section 2.3 by setting $\underline{B}_i = \underline{0}$, $i = 1, \ldots, p$:

$$\frac{\partial \underline{x}(t,\underline{z})}{\partial t} = \underline{A}\,\underline{x}(t,\underline{z}) + \underline{C}(\underline{z}) \cdot \underline{u}^{(1)}(t), \quad \underline{z}\in\Omega, \tag{4.27}$$

$$\underline{R}\,\underline{x}(t,\underline{z}) = \underline{D}(\underline{z}) \cdot \underline{u}^{(2)}(t), \quad \underline{z}\in\Gamma, \tag{4.28}$$

$$\underline{x}(0,\underline{z}) = \underline{x}_o(z), \quad \underline{z}\in\Omega. \tag{4.29}$$

$$\underline{y}(t) = \underline{E}\,\underline{x}(t,\underline{z}), \tag{4.30}$$

all operators being defined as in section 2.3.

Again by introducing the total state vector

$$\underline{\eta}(t,\underline{z}) = \begin{bmatrix} \underline{\xi}(t) \\ \underline{x}(t,\underline{z}) \end{bmatrix}, \tag{4.31}$$

Eqs.(4.24) - (4.30) can be summarized as

$$\frac{\partial \underline{\eta}}{\partial t} = \begin{bmatrix} \underline{\tilde{A}} & - \underline{\tilde{B}}\,\underline{K}(t)\underline{E} \\ \underline{C}(z)\underline{\tilde{C}}^{(1)} & \underline{A} \end{bmatrix} \cdot \underline{\eta}, \quad \underline{z}\in\Omega, \tag{4.32}$$

$$\underline{\eta}(0,\underline{z}) = \underline{\eta}_o(\underline{z}) = \begin{bmatrix} \underline{\xi}_o \\ \underline{x}_o(\underline{z}) \end{bmatrix}, \qquad \underline{z}\in\Omega, \tag{4.33}$$

$$\underline{R}\,\underline{x} = \underline{D}(\underline{z})\cdot\underline{\tilde{C}}^{(2)}\underline{\xi}, \qquad \underline{z}\in\Gamma. \tag{4.34}$$

In Eq.(4.32) $\underline{K}(t)$ plays the role of a matrix-type multiplicative control function. In most situations, a suitable strategy for the switching between two pre-specified matrices $\underline{K}_1$ and $\underline{K}_2$ will already bring about the desired effect [2], hence

$$\underline{K}(t) = \frac{1}{2}(\underline{K}_1+\underline{K}_2) + \frac{1}{2}(\underline{K}_1-\underline{K}_2)\tilde{u}(t), \tag{4.35}$$

with *scalar* control variable $\tilde{u}(t)$ taking only values + 1 and - 1:

$$\underline{K}(t) = \begin{cases} \underline{K}_1 & \text{if } \tilde{u} = +1, \\ \underline{K}_2 & \text{if } \tilde{u} = -1. \end{cases} \tag{4.36}$$

It should be noted, that the above over-all system is homogeneous-in-the-state. Therefore, if the zero state is asymptotically stable for every $\underline{K}(t)$ according to Eqs.(4.35), (4.36) and for all $\underline{\eta}_o(\underline{z})$, then the system will not be completely controllable. But the primary objective here is the design of switching strategies which make $\underline{y}(t)$ approach $\underline{0}$ as fast as possible (section 6.3).

4.2.2 An integral model

Integral models are often advantageous, as they include the influence of boundary and initial conditions. Moreover, the dimension of the vector-valued integral equation for the output of the over-all system can in most situations be made considerably smaller than the dimension of the total state vector [54].

Taking Eqs.(4.24) - (4.30) as a basis and using notations of Eq.(2.39), the linear distributed part can now be written as

$$\underline{x}(t,\underline{z}) = \int_{\tau=0}^{t}\left[\int_{\Omega}\underline{G}(t-\tau,\underline{z},\underline{\zeta})\underline{C}(\underline{\zeta})\,d\Omega_{\zeta}\right]\underline{u}^{(1)}(\tau)\,d\tau +$$

$$+ \int_{\tau=0}^{t} \left[\oint_{\Gamma} \underline{L}\{\underline{G}(t-\tau,\underline{z},\underline{\zeta})\}\underline{D}(\underline{\zeta})\,d\Gamma_{\underline{\zeta}} \right] \underline{u}^{(2)}(\tau)\,d\tau \; +$$

$$+ \int_{\Omega} \underline{G}(t,\underline{z},\underline{\zeta})\,\underline{x}_o(\underline{\zeta})\,d\Omega_{\underline{\zeta}} , \tag{4.37}$$

or, by means of abbreviations

$$\underline{G}^{(1)}(t-\tau,\underline{z}) = \int_{\Omega} \underline{G}(t-\tau,\underline{z},\underline{\zeta})\,\underline{C}(\underline{\zeta})\,d\Omega_{\underline{\zeta}} , \tag{4.38}$$

$$\underline{G}^{(2)}(t-\tau,\underline{z}) = \oint_{\Gamma} \underline{L}\{\underline{G}(t-\tau,\underline{z},\underline{\zeta})\}\underline{D}(\underline{\zeta})\,d\Gamma_{\underline{\zeta}} , \tag{4.39}$$

$$\underline{G}^{*}(t-\tau,\underline{z}) = [\underline{G}^{(1)}(t-\tau,\underline{z}),\ \underline{G}^{(2)}(t-\tau,\underline{z})] : \tag{4.40}$$

$$\underline{x}(t,\underline{z}) = \int_{\tau=0}^{t} \underline{G}^{*}(t-\tau,\underline{z})\,\underline{u}(\tau)\,d\tau + \int_{\Omega} \underline{G}(t,\underline{z},\underline{\zeta})\,\underline{x}_o(\underline{\zeta})\,d\Omega_{\underline{\zeta}} . \tag{4.41}$$

In the following, $\underline{y}(t)$ may be given by pointwise measurement,

$$y_i(t) = \underline{c}_i^{*T}\underline{x}(t,\underline{z}_i),\quad i = 1, \ldots, q, \tag{4.42}$$

$\underline{c}_i^{*T}$ and $\underline{z}_i$ denoting specified observation row vectors and measurement points, respectively. Then, by rewriting Eq.(4.41) for $\underline{z} = \underline{z}_i$, $i = 1, \ldots, q$, the following simple equation is obtained for $\underline{y}(t)$:

$$\underline{y}(t) = \int_{\tau=0}^{t} \underline{H}^{*}(t-\tau)\,\underline{u}(\tau)\,d\tau + \int_{\Omega} \underline{H}(t,\underline{\zeta})\,\underline{x}_o(\underline{\zeta})\,d\Omega_{\underline{\zeta}} , \tag{4.43}$$

where

$$\underline{H}^{*}(t-\tau) = \begin{bmatrix} \underline{c}_1^{*T}\underline{G}^{*}(t-\tau,\underline{z}_1) \\ \vdots \\ \underline{c}_q^{*T}\underline{G}^{*}(t-\tau,\underline{z}_q) \end{bmatrix} ,\qquad \underline{H}(t,\underline{\zeta}) = \begin{bmatrix} \underline{c}_1^{*T}\underline{G}(t,\underline{z}_1,\underline{\zeta}) \\ \vdots \\ \underline{c}_q^{*T}\underline{G}(t,\underline{z}_q,\underline{\zeta}) \end{bmatrix} . \tag{4.44}$$

Next, the lumped part will be written in integral form as well. Therefore, let $\underline{\Phi}(t) = \exp(\tilde{\underline{A}}t)$ be the transition matrix corresponding to $\tilde{\underline{A}}$. Then

$$\underline{u}(\tau) = \tilde{\underline{C}}\,\underline{\xi}(\tau) = \tilde{\underline{C}} \int_{0}^{\tau} \underline{\Phi}(\tau-\tau^{*})\,\tilde{\underline{B}}[-\underline{K}(\tau^{*})\,\underline{y}(\tau^{*})]\,d\tau^{*} + \tilde{\underline{C}}\,\underline{\Phi}(\tau)\,\underline{\xi}_o . \tag{4.45}$$

Now by inserting Eq.(4.45) into Eq.(4.43) the following integral equation is obtained for $\underline{y}(t)$:

$$\underline{y}(t) = \int_{\tau=0}^{t} \underline{\tilde{G}}(t-\tau)\underline{K}(\tau)\underline{y}(\tau)d\tau + \underline{y}_o(t), \tag{4.46}$$

where

$$\underline{y}_o(t) = \int_{\tau=0}^{t} \underline{H}^*(t-\tau)\underline{\tilde{C}}\,\underline{\Phi}\,(\tau)d\tau\cdot\underline{\xi}_o + \int_{\Omega}\underline{H}(t,\underline{\zeta})\underline{x}_o(\underline{\zeta})d\Omega_{\underline{\zeta}} \tag{4.47}$$

is caused by initial perturbation $\underline{\eta}_o(\underline{z})$, and the matrix kernel $\underline{\tilde{G}}(t-\tau)$ is obtained after some transformations as

$$\underline{\tilde{G}}(t-\tau) = -\int_{\tau^*=0}^{t-\tau} \underline{H}^*[(t-\tau)-\tau^*]\underline{\tilde{C}}\,\underline{\Phi}\,(\tau^*)\underline{\tilde{B}}d\tau^*. \tag{4.48}$$

Due to finite-dimensional output vector $\underline{y}(t)$, Eq.(4.46) for the over-all system formally seems to represent a bilinear *lumped* parameter system. But one should be aware of the fact that the elements of $\underline{\tilde{G}}(t-\tau)$ will *not* be composed by a *finite* number of exponential functions.

Nevertheless, the compact form of Eq.(4.46) proves to be a suitable basis for the design of optimal or near-optimal control. Fig.4.5 outlines the fundamental structure of Eq.(4.46), any switching strategy being omitted. The structure reminds to what is usually referred to as a time-varying system [55]. But the notion of a bilinear system seems to be more adequate here, as $\underline{K}(t)$ plays the role of a control input! The question of optimal switching strategy for this class of systems will be treated to some extent in chapter 6.

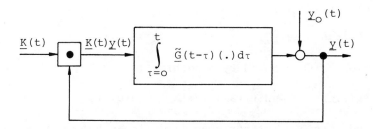

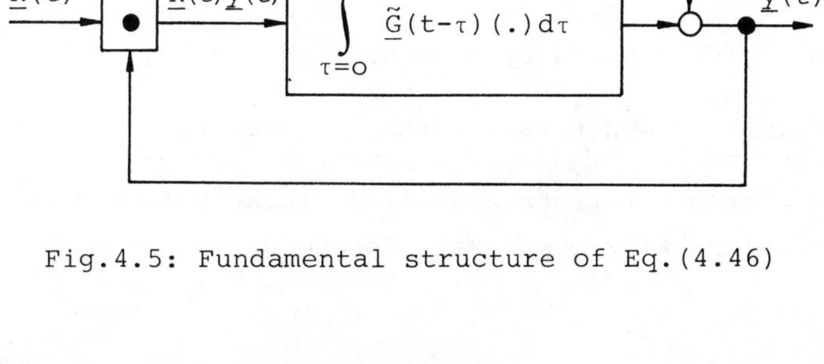

Fig.4.5: Fundamental structure of Eq.(4.46)

5. Design of closed-loop control via Ljapunow's direct method

In section 3.2 stability of a class of *open-loop* bilinear systems has been considered with no regard of the quality of stability or performance. But as Kalman and Bertram [45] pointed out in an early paper on this subject in the lumped parameter case, the degree of stability of a given steady state can considerably be improved by a possibly nonlinear feedback control law to be designed via Ljapunow's direct method.

5.1 Quasi-optimal nondynamic state feedback

For the ease of presentation the bilinear distributed system is assumed to be one-dimensional, with spatial coordinate $z\in[0,1]$. Both, control variable $u(t)$ and state variable $x(t,z)$ are assumed to be scalar. More complex situations can of course be treated by the same techniques.

According to whether $u(t)$ is acting in the interior of the system or on its boundary, the following models I) and II), in terms of perturbations, will be distinguished:

I) $$\frac{\partial x(t,z)}{\partial t} = Ax(t,z) + u(t)\cdot Bx(t,z) + C(z)\cdot u(t), \qquad 0\leq z\leq 1, \tag{5.1}$$

along with specified linear homogeneous boundary conditions, and initial perturbation

$$x(0,z) = x_o(z). \tag{5.2}$$

II) $$\frac{\partial x(t,z)}{\partial t} = Ax(t,z), \qquad 0\leq z\leq 1, \tag{5.3}$$

along with a boundary condition

$$Rx(t,z)\Big|_{z=0} = u(t)\cdot Bx(t,z)\Big|_{z=0} + d\cdot u(t), \tag{5.4}$$

and possibly further specified linear homogeneous boundary conditions at z=0 and z=1, depending on the type of the spatial differential operator A.

Initial perturbation

$$x(0,z) = x_o(z). \tag{5.5}$$

For both, models I) and II), the following additional assumptions will be made:

(a) The set of admissible controls, S, is given by

$$S = \{u(t) \mid m \leq u(t) \leq M\}. \tag{5.6}$$

Of course, S contains the origin as in terms of perturbations u=0 corresponds to the equilibrium control u_s of the original system.

(b) The zero steady state of the linear subsystem

$$\frac{\partial x(t,z)}{\partial t} = Ax(t,z), \qquad 0 \leq z \leq 1, \tag{5.7}$$

along with specified boundary conditions in their homogeneous form, is assumed to be asymptotically stable with respect to metric

$$\rho(x,0) = \|x\| = \sqrt{\int_0^1 x^2(t,z)dz}. \tag{5.8}$$

Now the following problem of quasi-optimal feedback control will be considered, rather than to pre-specify a definite control law with merely some free parameters.

Define a suitable Ljapunow-functional for the above models I) or II), e.g.

$$V = \frac{1}{2}\int_{z=0}^{1} x^2(t,z)dz, \tag{5.9}$$

and determine $u(t) \in S$ such as to minimize, in addition to asymptotic stability, the time derivative $\dot{V}$:

$$\dot{V} \stackrel{!}{=} \min_{u \in S}. \tag{5.10}$$

<u>Remark 1</u>: Note that due to assumption (b) $\dot{V}$ can in any case be made negative by simply setting $u(t) \equiv 0$. The design objective

is: *exhaust* the admissible set S such as to make $\dot{V}$ as negative as possible.

Remark 2: As in the linear lumped parameter case [45] the general inequality

$$\min_{u \in S} \frac{V(x,u)}{-\dot{V}(x,u)} < V(x,u) \cdot \min_{u \in S} \frac{1}{-\dot{V}(x,u)} \tag{5.11}$$

holds which makes obvious that this is not a method of true optimal control. Only quasi-optimal control in the sense of Eq.(5.10) can be achieved.

The procedure of minimizing $\dot{V}$ with respect to u is very simple and straight forward:

According to Eq.(5.9), we obtain

$$\dot{V} = \int_{z=0}^{1} x \frac{\partial x}{\partial t} dz, \tag{5.12}$$

hence in problem I):

$$\dot{V} = \int_0^1 x \cdot Axdz + u(t) \cdot \left[\int_0^1 x \cdot Bxdz + \int_0^1 C(z) \cdot xdz \right], \tag{5.13}$$

in problem II):

$$\dot{V} = \int_0^1 x \cdot Axdz. \tag{5.14}$$

The boundary condition (5.4) can be introduced into Eq.(5.14) via integration by parts. Therefore, the expression obtained for $\dot{V}$ will in both cases be *linear with respect to* u *and quadratic with respect to* x.

Now, by continuing the exemplary discussion of problem I), it can be noticed that in view of assumption (b) the first integral on the right hand side of Eq.(5.13) will be negative. Moreover, by exploiting assumption (a), $\dot{V}$ can be made as negative as possible by specifying the feedback control law

$$u(t) = \frac{M+m}{2} - \frac{M-m}{2} \operatorname{sgn}\left[\int_0^1 x \cdot Bxdz + \int_0^1 C(z) \cdot xdz \right]. \tag{5.15}$$

Similar expressions will be obtained by minimizing $\dot{V}$ in problem II).

Obviously, the quasi-optimal feedback control law, Eq.(5.15), is of the bang-bang type including the possibility of singular control during some intervals of time. Singular control motions arise whenever

$$\int_0^1 x \cdot Bx dz + \int_0^1 C(z) \cdot x dz \equiv 0 \qquad (5.16)$$

during some interval which makes $\dot{V}$ not explicitly depend on u. The difficult question of existence of solutions in view of the discontinuous sgn-function can be avoided by replacing the sgn-function by an arbitrarily good continuous function, e.g. a saturation. This avoids at the same time sliding modes.

In any case, application of the above control law will considerably improve the performance of the controlled system compared to the uncontrolled one (see example 5.1).

It should be noted that Eq.(5.15) will, in general, require feedback of a nonlinear functional of the *total* state x(t,z) of the perturbed system. This will in most cases not be implementable unless state observers due to Köhne [46] and Zeitz [60] are taken into consideration. One should also be aware of additional problems of stability of the so obtained over-all system, which are by no means trivial.

On the other hand, state observers may be avoided by suitable approximations of the arising functionals. In many situations a finite number of measurement points z_i, i=1, ..., q, with measurements $x_i(t) = x(t,z_i)$ are available. The functionals arising in Eq.(5.15) may then be approximated by using straight lines or Simpson's formula (see example 5.1). It should be noticed that the use of the approximated control law will not assure stability of the controlled system. An expedient may be the application of this control law during a *finite time* and then passing over to $u \equiv 0$.

Example 5.1:

As an example of type I consider the process described by Eqs. (3.4) - (3.6) in example 3.1, with scalar control variable $u_1(t)$ and fixed $u_2 \equiv u_{2s}$, $u_3 \equiv u_{3s}$. Then the perturbed equations, according to example 3.3, assume the form

$$\frac{\partial x}{\partial t} = - cx(t,z) - u_{1s}\frac{\partial x}{\partial z} - u_1(t)\frac{\partial x}{\partial z} - x_s'(z)u_1(t), \qquad 0 \leqq z \leqq 1, \tag{5.17}$$

with homogeneous boundary condition

$$x(t,0) = 0 \tag{5.18}$$

and initial perturbation

$$x(0,z) = x_o(z). \tag{5.19}$$

The bounds on perturbation $u_1(t)$,

$$m_1 \leqq u_1(t) \leqq M_1, \tag{5.20}$$

are of course different from those in Eq.(3.6).

Now in terms of Eq.(5.1)

$$A = - cI - u_{1s}\frac{\partial}{\partial z}, \text{ (I identity operator)}, \tag{5.21}$$

$$B = - \frac{\partial}{\partial z}, \tag{5.22}$$

$$C(z) = - x_s'(z) = \frac{c}{u_{1s}}(u_{3s}-u_{2s})\exp(- \frac{c}{u_{1s}} z). \tag{5.23}$$

The latter expression has been obtained from Eq.(3.7).

By specifying the Ljapunow-functional

$$V = \frac{1}{2}\int_0^1 x^2(t,z)dz,$$

and by means of some simple calculations one obtains

$$\dot{V} = - c\int_0^1 x^2(t,z)dz - \frac{u_{1s}}{2} x^2(t,1) -$$

$$- u_1(t) \cdot \left[\frac{1}{2} x^2(t,1) + \int_0^1 x_s'(z) x(t,z) dz \right] , \tag{5.24}$$

which corresponds to Eq.(5.13). Of course, the first two terms on the right hand side of Eq.(5.24) are $\leqq 0$ and vanish only in the zero equilibrium state thus indicating asymptotic stability of the uncontrolled system. Hence assumption (b) holds which can also be concluded for physical reasons in this very simple process.

Application of Eq.(5.15) now immediately yields the quasi-optimal control law

$$u_1(t) = \frac{M_1+m_1}{2} + \frac{M_1-m_1}{2} \mathrm{sgn} \left[\frac{1}{2} x^2(t,1) + \int_0^1 x_s'(z) x(t,z) dz \right] , \tag{5.25}$$

which makes $\dot{V}$ as negative as possible by exhausting the admissible set S.

The following numerical example has been simulated by analog computation.

$$c = u_{1s} = u_{3s} = 1; \quad u_{2s} = 0;$$

$$m_1 = -1, \; M_1 = +1 \text{ in Eq.(5.20)},$$

hence perturbation $u_1 = -1$ means standstill of the conveyor belt.

As now $x_s(z) = \exp(-z)$ and $x_s'(z) = -\exp(-z)$, the control law (5.25) takes the form

$$u_1(t) = \mathrm{sgn} \left[\frac{1}{2} x^2(t,1) - \int_0^1 e^{-z} x(t,z) dz \right] . \tag{5.26}$$

The integral in Eq.(5.26) has been approximated by assuming equidistant measurement points $z_i = i/5$; $i=1, \ldots, 5$, and by using straight line approximation to obtain

$$u_1(t) \approx \mathrm{sgn}[- 0{,}328 x_1(t) - 0{,}268 x_2(t) - 0{,}22 x_3(t) -$$
$$- 0{,}18 x_4(t) - 0{,}074 x_5(t) + x_5^2(t)]. \tag{5.27}$$

The bilinear process itself has been simulated using spatial discretization, with the same grid points z_i as denoted above. This fifth order lumped approximation provides, of course, only a very rough idea of the system behaviour. Nevertheless, the fundamental improvement achieved by means of the above control law can be seen from Figures 5.1 and 5.2.

Fig. 5.1 illustrates performance of the uncontrolled simulated system ($u_1 \equiv 0$) with initial disturbances $x(0,z_i)=x_0=0{,}5$, $i=1, \ldots, 5$. Control law (5.27) has been employed to this system to obtain Fig. 5.2. Obviously, the integral of the square error,

$$\sum_{i=1}^{5} \int_{t=0}^{\infty} x_i^2(t)\,dt \approx 10 \int_{t=0}^{\infty} V dt,$$

has been reduced to a significant amount.

It can be observed that only for $0 \leqq t \leqq 0{,}6$ the control is bang-bang type. For $t>0{,}6$ the controlled system works in a sliding

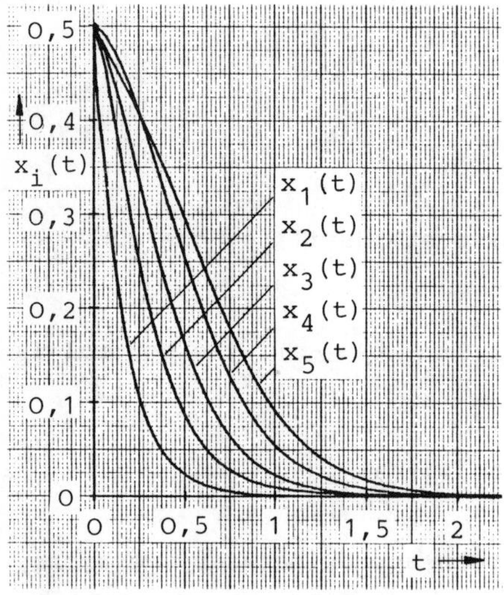

Fig.5.1: Performance of the uncontrolled simulated system; $x_i(t) = x(t,\frac{i}{5})$

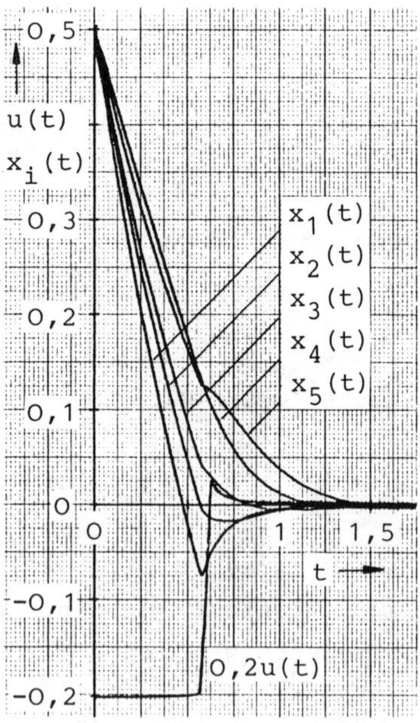

Fig.5.2: Performance with control law (5.27)

mode, which can be avoided by replacing the sgn-function by a continuous function.

In Figures 5.3 and 5.4 the behaviour of the uncontrolled and controlled system is graphed, respectively, for the case of initial disturbances

$$x_1(0) = x_2(0) = x_5(0) = +0{,}5;$$

$$x_3(0) = x_4(0) = -0{,}5.$$

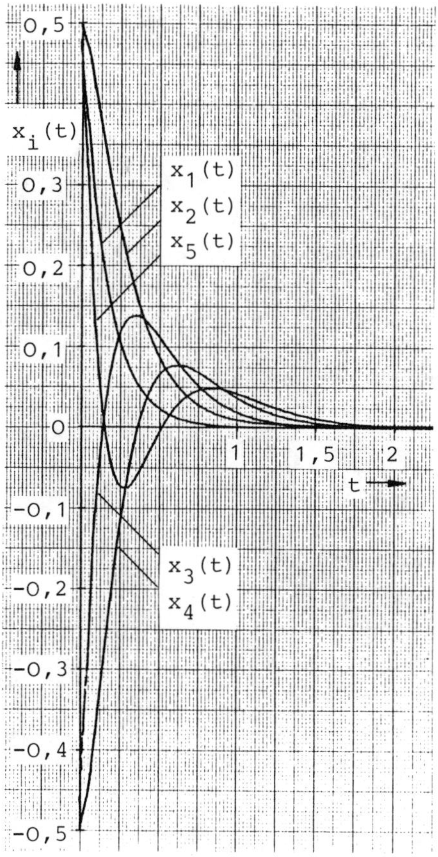

Fig.5.3: Performance of the uncontrolled simulated system

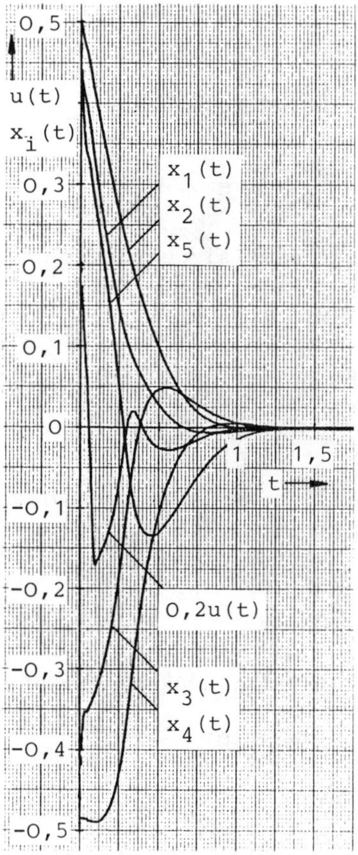

Fig.5.4: Performance with control law (5.27)

Again, the performance of the controlled system can be observed to be better to a considerable degree. Here, the system works in a sliding mode from the beginning.

Example 5.2:

This is an example of type II) with a multiplicative boundary control. Example 2.3 will be considered with scalar control variable $u_2(t)$, $u_{2s}>0$, and fixed $u_1\equiv 0$, $u_3\equiv u_{3s}\not\equiv 0$. Then the perturbed equations, with regard to example 3.4, assume the form

$$\frac{\partial x}{\partial t} = a\frac{\partial^2 x}{\partial z^2}, \quad 0\leqq z\leqq 1, \tag{5.28}$$

$$\left[\frac{\partial x}{\partial z} - u_{2s}x(t,z)\right]_{z=0} = u_2(t)x(t,0) + x_s(0)u_2(t), \tag{5.29}$$

$$\left.\frac{\partial x}{\partial z}\right|_{z=1} = 0, \tag{5.30}$$

with initial perturbation

$$x(0,z) = x_o(z). \tag{5.31}$$

The set of admissible control perturbations $u_2(t)$,

$$m_2\leqq u_2(t)\leqq M_2, \tag{5.32}$$

will in any case contain the origin.

In virtue of notations in Eqs.(5.3) and (5.4) now

$$A = a\cdot\frac{\partial^2}{\partial z^2}, \tag{5.33}$$

$$B = I \text{ (identity operator)}, \tag{5.34}$$

$$d = x_s(0) = u_{3s}/u_{2s} \tag{5.35}$$

Again

$$V = \frac{1}{2}\int_0^1 x^2(t,z)\,dz$$

will be chosen as a Ljapunow-functional, hence

$$\dot{V} = \int_0^1 x \frac{\partial x}{\partial t} dz = a \int_0^1 x \frac{\partial^2 x}{\partial z^2} dz.$$

Integration by parts and inserting the boundary conditions (5.29), (5.30) finally yields

$$\dot{V} = - a u_{2s} x^2(t,0) - a \int_0^1 \left(\frac{\partial x}{\partial z}\right)^2 dz -$$

$$- a u_2(t) x(t,0) \cdot [x(t,0) + u_{3s}/u_{2s}]. \tag{5.36}$$

Similar to example 5.1, Eq.(5.24), it can be noticed that the first two terms on the right hand side of Eq.(5.36) are ≤ 0 and vanish only in the zero equilibrium state thus making assumption (b) valid (see also example 3.6).

The quasi-optimal control law according to Eq.(5.15) is now obtained as

$$u_2(t) = \frac{M_2+m_2}{2} + \frac{M_2-m_2}{2} \operatorname{sgn} x(t,0) \cdot \operatorname{sgn} \left[x(t,0) + \frac{u_{3s}}{u_{2s}} \right]. \tag{5.37}$$

Obviously, no functional must be calculated in this example. Moreover, the surface temperature $x(t,0)$ will be measurable in most situations which makes any approximation dispensable.

In those situations where the original boundary temperature

$$x(t,0) + u_{3s}/u_{2s} > 0$$

for physical reasons, the above control law can be simplified to

$$u_2(t) = \frac{M_2+m_2}{2} + \frac{M_2-m_2}{2} \operatorname{sgn} x(t,0). \tag{5.38}$$

5.2 Design of linear lumped dynamic feedback

No dynamical subsystems, in addition to the bilinear distributed process, have been considered in section 5.1. But in most cases sensors, actuators and controllers must be considered to be dynamical systems as well. If, by ignoring them, the control design proposed in section 5.1 were applied, stability of the actual over-all system can of course not be assured. Moreover, the bang-bang controller along with certain low pass nature of the dynamical part of the system will, in general, give rise to limit cycles of the controlled system which have to be avoided. The describing function method may be an appropriate tool of handling this problem, but will not be discussed in this context.

An extension of Ljapunow's direct method to mixed lumped and distributed parameter systems has been proposed by Wang [56]. This technique proves to be feasible also for the problem in hand.

The following considerations are basing on the feedback control structure as described in section 4.1 and outlined in Fig. 4.1. As in section 5.1, the bilinear distributed subsystem is assumed to be one-dimensional.

Control variable u(t), state variable x(t,z) and output y(t), hence w and $y_e(t)$, are assumed to be scalar. u(t) is exemplary assumed to act in the interior of the system, but boundary control can be treated by the same techniques (see example 5.4).

Then the state equations of the closed-loop system outlined in Fig. 4.1, rewritten in terms of perturbations about a predetermined steady state (section 4.1.2), take the following form:

m - th order lumped subsystem:

$$\dot{\underline{\xi}} = \underline{\tilde{A}}\,\underline{\xi} - \underline{\tilde{b}}y, \tag{5.39}$$

$$u = \underline{\tilde{c}}^T \underline{\xi}, \tag{5.40}$$

$$\underline{\xi}(0) = \underline{\xi}_o. \tag{5.41}$$

Without loss of generality, $\underline{\tilde{c}}^T = \underline{e}_m^T = (0, \ldots, 0, 1)$ will subsequently be assumed, hence

$$u = \underline{e}_m^T \underline{\xi} = \xi_m. \tag{5.42}$$

Distributed subsystem:

$$\frac{\partial x}{\partial t} = Ax + u \cdot Bx + C(z) \cdot u, \qquad 0 \leqq z \leqq 1, \tag{5.43}$$

$$Rx = 0, \qquad z = 0,1; \tag{5.44}$$

$$x(0,z) = x_o(z). \tag{5.45}$$

The output equation will alternatively be specified by

$$y(t) = Ex(t,z) = \int_{z=0}^{1} \delta(z,z_M)x(t,z)dz = x(t,z_M), \tag{5.46}$$

or, with regard to Eq.(5.15), by

$$y(t) = Ex(t,z) = \int_{z=0}^{1} x \cdot Bxdz + \int_{z=0}^{1} C(z) \cdot xdz. \tag{5.47}$$

Eq.(5.46) means pointwise measurement, with specified measurement point z_M; Eq.(5.47) means that a weakened version of the bang-bang type control of section 5.1 is taken into consideration.

With respect to the controller design, there will be at least one free parameter, k, in the equations of the lumped subsystem. As a realistic example, it will be assumed in the subsequence that k is the gain of the controller, hence

$$\underline{\tilde{b}} = k\underline{b}^*, \tag{5.48}$$

where $\underline{b}^*$ and $\underline{\tilde{A}}$ are specified.

Therefore, different from the considerations in section 5.1, the problem of stability of the zero equilibrium state is now a *parameter*-problem, with parameter k to be specified.

The following additional assumptions will be made:

(a), (b) } as in section (5.1).

(c) The interval [m,M], for the admissible control u(t), contains the zero in its *interior*.

(d) The linear lumped subsystem is assumed to be asymptotically stable, which requires the real parts of all eigenvalues of the (m,m)-matrix $\underline{\tilde{A}}$ to be negative.

Design of parameter k along with assurance of a domain of asymptotic stability will be treated for both, output equation (5.46) and (5.47), in sections 5.2.1 and 5.2.2, respectively.

5.2.1 Single-loop feedback control

By means of output equation (5.46) and by using the total (m+1)-dimensional state vector

$$\underline{\eta}(t,z) = \begin{bmatrix} \underline{\xi}(t) \\ x(t,z) \end{bmatrix} , \qquad (5.49)$$

the state equations of the autonomous closed-loop system can be rewritten in the form

$$\frac{\partial \underline{\eta}}{\partial t} = \begin{bmatrix} \underline{\tilde{A}} & -\underline{\tilde{b}}E \\ C(z)\underline{e}_m^T & A+\xi_m B \end{bmatrix} \underline{\eta} , \quad z\in[0,1]; \qquad (5.50)$$

$$Rx = 0, \; z = 0,1; \qquad (5.51)$$

$$\underline{\eta}(0,\underline{z}) = \underline{\eta}_o(z), \qquad (5.52)$$

with steady state $\underline{\eta}_s = \underline{0}$, whose stability behaviour is to be investigated.

Therefore the augmented Ljapunow-functional [24]

$$V = \underline{\xi}^T \underline{P}\, \underline{\xi} + \frac{1}{2}\int_0^1 x^2 dz \qquad (5.53)$$

is employed with yet undetermined symmetric (m,m)-matrix $\underline{P}$. V will be positive definite if and only if the quadratic form in $\underline{P}$ is positive definite.

Now

$$\dot{V} = \underline{\dot{\xi}}^T \underline{P}\, \underline{\xi} + \underline{\xi}^T \underline{P}\, \underline{\dot{\xi}} + \int_0^1 x \frac{\partial x}{\partial t} dz =$$

$$= \underline{\xi}^T (\underline{\tilde{A}}^T \underline{P} + \underline{P}\,\underline{\tilde{A}})\underline{\xi} - x(t,z_M)\underline{\tilde{b}}^T \underline{P}\,\underline{\xi} -$$

$$- \underline{\xi}^T \underline{P}\,\underline{\tilde{b}}\, x(t,z_M) +$$

$$+ \int_0^1 x\cdot Axdz + \xi_m \int_0^1 x\cdot Bxdz + \xi_m \int_0^1 C(z)xdz. \qquad (5.54)$$

At the first glance the determination of conditions for negative definiteness of $\dot{V}$ seems to be a severe problem in view of the very complex expression. But in many situations the sum S_1 of integral terms in Eq.(5.54),

$$S_1 = \int_0^1 x \cdot Axdz + \xi_m \int_0^1 x \cdot Bxdz + \xi_m \int_0^1 C(z)xdz, \qquad (5.55)$$

can be estimated by an expression of the following type which is free of any integral:

$$S_1 \leq (a_o + a_1\xi_m)x^2(t,z_M) + a_2\xi_m^2 + 2(a_3+a_4\xi_m)\xi_m x(t,z_M) =$$

$$= \underline{\eta}^T(t,z_M) \begin{bmatrix} a_2 \operatorname{diag}(\delta_{im}) & (a_3+a_4\xi_m)\underline{e}_m \\ (a_3+a_4\xi_m)\underline{e}_m^T & a_o+a_1\xi_m \end{bmatrix} \underline{\eta}(t,z_M), \qquad (5.56)$$

where

$$\delta_{im} = \begin{cases} 0 \text{ for } i \neq m \\ 1 \text{ for } i = m \end{cases} \qquad \text{(Kronecker's delta)},$$

and $a_o, \ldots, a_4$ are certain coefficients. a_o will in any case be negative due to assumption (b).

By inserting Eq.(5.56) into Eq.(5.54), the following quadratic form in $\underline{\eta}(t,z_M)$ is readily obtained as an upper bound of $\dot{V}$:

$$\dot{V} \leq \underline{\eta}^T(t,z_M) \begin{bmatrix} \underline{\tilde{A}}^T\underline{P}+\underline{P}\,\underline{\tilde{A}} + a_2\operatorname{diag}(\delta_{im}) & (a_3+a_4\xi_m)\underline{e}_m - \underline{P}\,\underline{\tilde{b}} \\ (a_3+a_4\xi_m)\underline{e}_m^T - \underline{\tilde{b}}^T\underline{P} & a_o+a_1\xi_m \end{bmatrix} \underline{\eta}(t,z_M) =$$

$$= \underline{\eta}^T(t,z_M)\underline{Q}\,\underline{\eta}(t,z_M) := \dot{V}_{max}. \qquad (5.57)$$

Of course, negative definiteness of this quadratic form can easily be examined by simply applying Sylvester's criterion [41], [48] to the matrix $-\underline{Q}$.

In this context it is important to keep in mind Eq.(5.48) which makes the matrix $\underline{Q}$ in Eq.(5.57) a function of only two scalar variables, k and ξ_m:

$$\underline{Q} = \underline{Q}(k,\xi_m). \tag{5.58}$$

Hence, the domain of negative definite $\underline{Q}$, if it exists, is a domain in the plain spanned by the controller gain k and by the control variable $u = \xi_m$. An example is outlined in Fig.5.5.

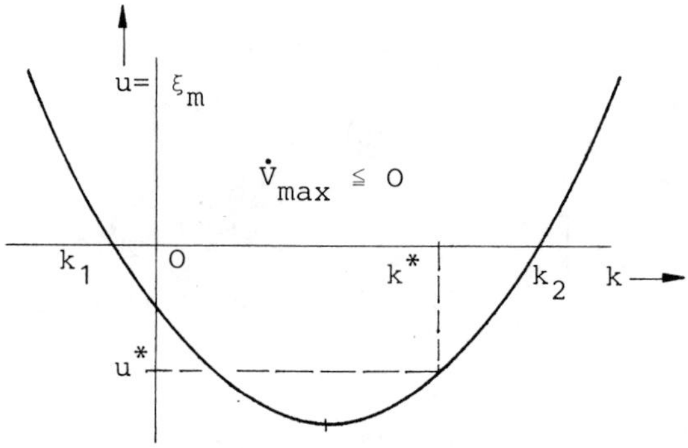

Fig.5.5: Example of the domain of negative definite $\underline{\eta}^T(t,z_M)\underline{Q}(k,u)\underline{\eta}(t,z_M)$

By specifying $k = k^*$, $\dot{V} \leqq 0$ is assured for all $u(t) \geqq u^*$. Here k^* must be specified between k_1 and k_2, as the domain of definiteness must contain the zero state!

It must be stressed that Eq.(5.57) is a quadratic form in $\underline{\eta}(t,z_M)$ and not in $\underline{\eta}(t,z)$. Therefore, negative definite $\underline{Q}$ makes $\dot{V}_{max}$ merely negative *semi*definite with regard to the over-all system. It must therefore be examined [56], if $\dot{V}_{max} \not\equiv 0$ along any perturbed motion, in which case asymptotic stability can be assured.

In detail, $\dot{V}_{max} \equiv 0$ if $\underline{\eta}(t,z_M) \equiv 0$. Therefore, by inserting $x(t,z_M) \equiv 0$ and $\underline{\xi}(t) \equiv \underline{0}$, hence $u(t) = \xi_m(t) \equiv 0$, into the state equations, the system's motions are obtained along which $\dot{V}_{max} \equiv 0$:

$$\text{Eq.}(5.39) \wedge \dot{\underline{\xi}}(t) \equiv \underline{0};$$

$$\text{Eq.}(5.43) \wedge \frac{\partial x}{\partial t} = Ax. \tag{5.59}$$

Now if Eq.(5.59) along with both,

$$Rx = 0 \quad \textit{and} \quad x(t,z_M) = 0, \tag{5.60}$$

and initial condition

$$x(0,z) = x_o(z) \tag{5.61}$$

has the only solution $x(t,z) \equiv 0$ which implies $x_o(z) \equiv 0$, then $\dot{V}_{max} \not\equiv 0$ along any perturbed motion of the system under consideration. In this case V with positive definite $\underline{P}$ is a suitable Ljapunow-functional.

Now, keeping in mind this statement, Sylvester's criterion has to be applied to the matrix

$$-\underline{Q}(k,u) = \begin{bmatrix} -\underline{\tilde{A}}^T\underline{P}-\underline{P}\,\underline{\tilde{A}}-a_2\mathrm{diag}(\delta_{im}) & k\underline{P}\,\underline{b}^*-(a_3+a_4u)\underline{e}_m \\ k\underline{b}^{*T}\underline{P}-(a_3+a_4u)\underline{e}_m^T & -a_o-a_1u \end{bmatrix}, \tag{5.62}$$

with yet unspecified $\underline{P}$.

Now the determination of $\underline{P}$ from the Ljapunow equation

$$\underline{\tilde{A}}^T\underline{P} + \underline{P}\,\underline{\tilde{A}} = -\gamma\underline{I}, \tag{5.63}$$

with real number γ satisfying

$$\gamma > \max\{0,a_2\}, \tag{5.64}$$

assures positive definiteness of $\underline{P}$ in view of assumption (d) on the one hand, and makes the upper left-hand corner of $-\underline{Q}$,

$$-\underline{\tilde{A}}^T\underline{P}-\underline{P}\,\underline{\tilde{A}}-a_2\mathrm{diag}(\delta_{im}) = \gamma\cdot\underline{I}-a_2\mathrm{diag}(\delta_{im}),$$

positive definite on the other hand. Hence, only the determinant

$$\det[-\underline{Q}(k,u)] = \begin{vmatrix} \gamma\underline{I}-a_2\mathrm{diag}(\delta_{im}) & k\underline{P}\,\underline{b}^*-(a_3+a_4u)\underline{e}_m \\ k\underline{b}^{*T}\underline{P}-(a_3+a_4u)\underline{e}_m^T & -a_o-a_1u \end{vmatrix} \tag{5.65}$$

is crucial in the application of Sylvester's criterion.

Keeping in mind that due to assumption (b) a_o will in any case be negative, Sylvester's inequality

$$\det[-\underline{Q}(0,0)] = \begin{vmatrix} \gamma\underline{I}-a_2\mathrm{diag}(\delta_{im}) & -a_3\underline{e}_m \\ -a_3\underline{e}_m^T & |a_o| \end{vmatrix} > 0 \qquad (5.66)$$

will always hold for $k = 0$, $u = 0$ by choosing γ sufficiently large thus assuring $\dot{V}_{max} \leqq 0$ in a neighbourhood of the origin in the k, u-plane. The exact contour of this neighbourhood can be determined from Eq. (5.65) by setting

$$\det[-\underline{Q}(k,u)] = 0. \qquad (5.67)$$

It can easily be verified that the evaluation of this determinant leads to an equation of the type

$$\alpha_1 u+\alpha_2 u^2+\alpha_3 k+\alpha_4 k^2+\alpha_5 uk+\alpha_6 = 0, \qquad (5.68)$$

with certain coefficients $\alpha_1, \ldots, \alpha_6$. In the special case where $a_4 = 0$, a parabola is obtained ($\alpha_2 = \alpha_5 = 0$) as outlined in Fig. 5.5. If in addition $a_3 = 0$, the parabola is symmetric to the u-axis ($\alpha_3 = 0$). In any case, the domain where $\dot{V}_{max} \leqq 0$ contains the origin in its interior.

If now a fixed k is supposed to be specified such that

$$\dot{V}_{max} \leqq 0 \text{ for each } u\in U(k) := \{u(t)\,|\,m\leqq u^*(k)\leqq u(t)\leqq u^{**}(k)\leqq M\}, \qquad (5.69)$$

U containing the origin in its interior, then in terms of the total state of the system we have

$$\dot{V} \leqq 0 \text{ for each } \underline{\eta}(t,z)\in\Theta(k) := \{\underline{\eta}(t,z)\,|\,\xi_m(t)\in U(k)\}. \qquad (5.70)$$

Of course, Θ is not the domain of asymptotic stability to be assurable by Ljapunow's direct method. Only the subdomain $\Lambda\subset\Theta$ with property $V<\beta$, where β is a positive number, can be assured, if in addition $\dot{V}_{max} \not\equiv 0$ along any perturbed motion. Hence, the hyper-surfaces V = const. have to be determined:

$$V = \underline{\xi}^T\underline{P}\,\underline{\xi} + \frac{1}{2}\int_0^1 x^2(t,z)\,dz =$$

$$= \underline{\xi}^T \underline{P}\, \underline{\xi} + \frac{1}{2} ||x||^2 = \beta, \qquad (5.71)$$

with L_2-norm

$$||x|| = \sqrt{\int_0^1 x^2(t,z)\,dz}. \qquad (5.72)$$

Eq.(5.71) describes a hyper-surface in an (m+1)-dimensional Euclidean space spanned by $\xi_1, \ldots, \xi_m, ||x||$. Due to $||x|| \geqq 0$, *half-ellipsoids* are obtained.

Now choose β as large as possible such that the half-ellipsoid, Eq.(5.71), lies entirely in the interior of the strip defined by $\xi_m \in U(k)$. Then the interior of this half-ellipsoid, Λ, is assured to belong to the domain of asymptotic stability. This completes the procedure of the direct method.

It should be noted that there will be some best value of k with regard to Eqs.(5.6) and (5.69). This will, in general, suggest k to be specified such that $u^*(k) = m$ and/or $u^{**}(k) = M$.

One should also be aware of the degree of freedom in the choice of γ in Eq.(5.63). This can be utilized to make the domain in the k, u-plane where $\dot{V}_{max} \leqq 0$ as large as possible. The special matrix $-\gamma\underline{I}$ may also be replaced by a more general negative definite matrix for this purpose.

Example 5.3:

The process described in example 5.1 is considered again (Eqs.(5.17) - (5.23)), with scalar output $y(t) = x(t,1)$ and with linear lumped parameter dynamical feedback. As a numerical example let the lumped part be given by its transfer function

$$G(s) = k \cdot \frac{1}{1+3s+s^2}, \qquad (5.73)$$

with yet undetermined gain $k > 0$. Then a suitable state space notation is obtained as

$$\underline{\dot{\xi}} = \underbrace{\begin{bmatrix} 0 & -1 \\ 1 & -3 \end{bmatrix}}_{\underline{\tilde{A}}} \underline{\xi} + k \underbrace{\begin{bmatrix} 1 \\ 0 \end{bmatrix}}_{\underline{b}^*} (-y); \quad u_1 = \xi_2. \tag{5.74}$$

Assumption (d) is satisfied.

Now S_1, according Eq.(5.55), is given by

$$S_1 = - c\int_0^1 x^2 dz - u_{1s}\int_0^1 x \frac{\partial x}{\partial z} dz - \xi_2\int_0^1 x \frac{\partial x}{\partial z} dz -$$

$$- \xi_2 \int_0^1 x_s'(z) x dz.$$

As

$$\int_0^1 x \frac{\partial x}{\partial z} dz = \frac{1}{2} x^2(t,1),$$

S_1 can be estimated as

$$S_1 \leqq - \frac{1}{2}(u_{1s} + \xi_2) x^2(t,1) - c\int_0^1 x^2 dz +$$

$$+ |\xi_2| \int_0^1 |x_s'(z)| \cdot |x| dz.$$

Now by applying Schwartz's inequality to the latter integral and by quadratic completion one obtains

$$S_1 \leqq - \frac{1}{2}(u_{1s} + \xi_2) x^2(t,1) + \frac{\kappa}{4c} \xi_2^2 , \tag{5.75}$$

where

$$\kappa = \int_0^1 x_s'^2(z) dz. \tag{5.76}$$

This is the estimate required in Eq.(5.56). Here

$$a_o = - \frac{u_{1s}}{2} < 0; \; a_1 = - \frac{1}{2}; \; a_2 = \frac{\kappa}{4c} ; \; a_3 = a_4 = 0.$$

Next it has to be examined if $\dot{V} \not\equiv 0$ along any perturbed motion, according to Eqs.(5.59) - (5.61):

$$\frac{\partial x}{\partial t} = - cx - u_{1s} \frac{\partial x}{\partial z} ,$$

$$x(t,0) = 0 \quad \textit{and} \quad x(t,1) = 0,$$

$$x(0,z) = x_o(z).$$

It can be concluded by means of the method of characteristics or directly from the physical nature of the system, that $x(t,1) = 0$ requires $x_o(z) \equiv 0$ which assures $\dot{V} \not\equiv 0$ along any perturbed motion.

From Eq.(5.63) $\underline{P}$ is obtained as

$$\underline{P} = \frac{\gamma}{6} \cdot \begin{bmatrix} 11 & -3 \\ -3 & 2 \end{bmatrix} ,$$

hence

$$-\underline{Q}(k,u_1) = \begin{bmatrix} \gamma & 0 & \frac{11}{6}\gamma k \\ 0 & \gamma-a_2 & -\frac{1}{2}\gamma k \\ \frac{11}{6}\gamma k & -\frac{1}{2}\gamma k & \frac{1}{2}(u_{1s}+u_1) \end{bmatrix} .$$

Evaluation of Eq.(5.67) yields the parabola

$$\frac{1}{2}(\gamma-a_2)(u_{1s}+u_1)-\gamma(3{,}61\gamma-3{,}36a_2)k^2 = 0.$$

The parameter $\gamma = 1{,}264a_2$ has been determined such as to make the parabola as wide as possible.

As a numerical example of the plant, $u_{1s} = u_{3s} = c = 1$, $u_{2s}=0$, hence $x_{1s}(z) = \exp(-z)$, has been assumed, and perturbation $u_1(t)$ to be bounded by

$$|u_1(t)| \leqq 0{,}5.$$

Fig. 5.6 outlines the parabola in the k, u_1-plane obtained for this example. $k = 0{,}63$ has been specified from the requirement that the strip $|u_1| \leqq 0{,}5$ belong to the domain where $\dot{V} \leqq 0$.

The largest assurable domain of asymptotic stability, Λ, in the ξ_1, ξ_2, $||x||$ - space due to Eq.(5.71) is obtained for $\beta = 0{,}00675$ and is outlined in Fig. 5.7.

As a result, asymptotic stability of the origin can be assured for any initial disturbance which lies inside the domain graphed in Fig.5.7, assuring at the same time the bounds imposed on $u_1(t)$. In the special case, where $\xi_1(0)=\xi_2(0)=0$, the plant initial disturbance must satisfy

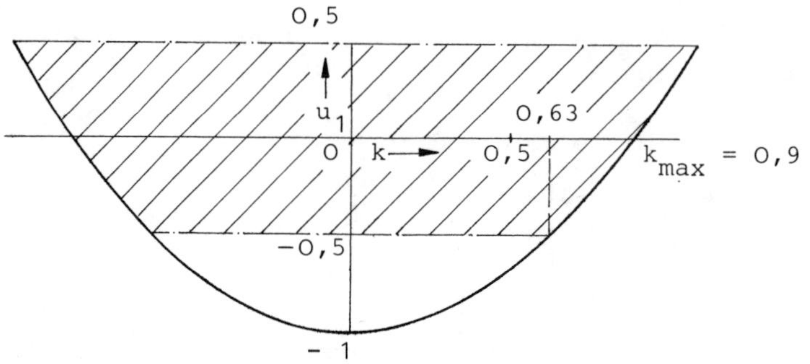

Fig.5.6: Domain of negative definite $\dot{V}_{max}$ in the k, u_1-plane; specification of k = 0,63 from bounded u_1

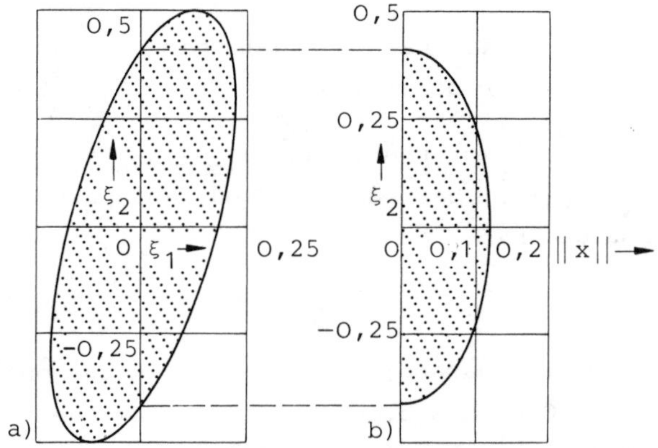

Fig.5.7: Domain of asymptotic stability in the ξ_1, ξ_2, $||x||$ -space;

a) Cross section in the ξ_1, ξ_2-plane

b) Cross section in the $||x||$, ξ_2-plane

$$||x_o(z)|| \leq 0,116.$$

The complete mixed lumped and distributed control system has been simulated, using the same method of approximation as in example 5.1 for the distributed part.

As can be seen from Fig.5.8, the controlled system is asymptotically stable even for the large initial disturbance $x_i(0) = x(0,i/5) = 0,5;\ i = 1, \dots, 5$. In Fig.5.9 the gain has been increased to $k = 5,7$, which still makes a stable over-all system. This is due to the fact that Ljapunow's direct method provides only sufficient conditions of stability, hence nothing can be said about the sharpness of the obtained conditions.

On the other hand, the procedure proposed is straight forward regardless of the order of the lumped part and without the need of approximating the distributed part.

Example 5.4:

Let the plant to be controlled the same as in example 5.2 with multiplicative boundary control variable $u_2(t)$ and output

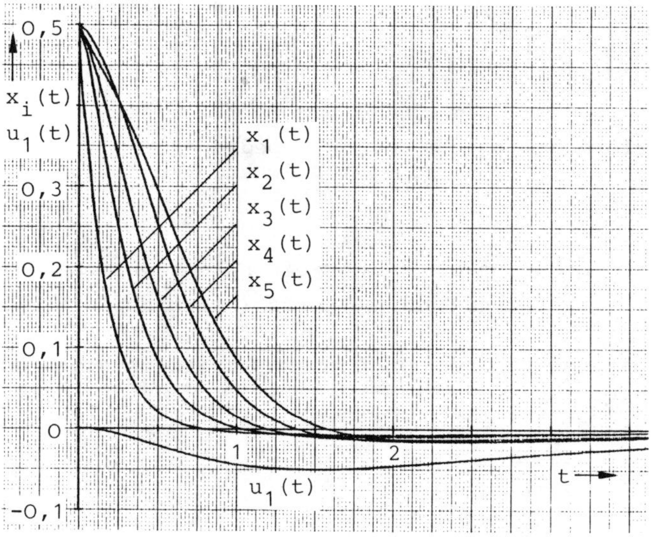

Fig.5.8: Performance of the mixed lumped and distributed control system, where $k = 0,63$.

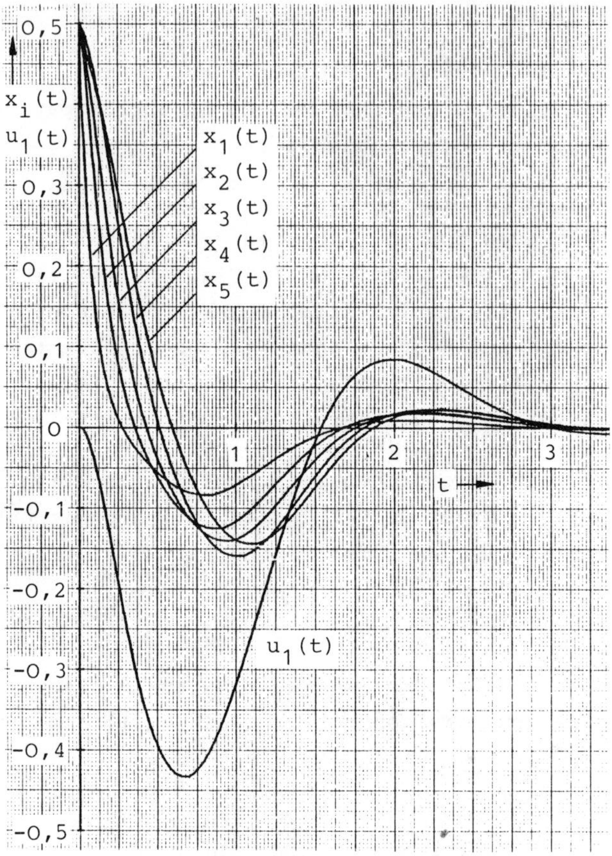

Fig.5.9: Performance of the mixed lumped and distributed control system, where k = 5,7.

$y(t) = x(t,0)$. Problems of this type can be transformed into previous ones via integration by parts (see example 5.2).

The lumped part of the control loop may again be given numerically as

$$G(s) = k \cdot \frac{1}{1+3s+s^2}$$

with yet undetermined k which, in view of example 4.2, will be required to be *negative* in this case.

State space notation (5.74) will be used again, assumption (d) being satisfied.

Now by choosing V as proposed in Eq.(5.53), we obtain

$$\dot{V} = \underline{\xi}^T(\underline{\tilde{A}}^T\underline{P} + \underline{P}\,\underline{\tilde{A}})\underline{\xi} - x(t,0)k\underline{b}^{*T}\underline{P}\,\underline{\xi} - - \underline{\xi}^T\underline{P}k\underline{b}^* x(t,0) + a\int_0^1 x\,\frac{\partial^2 x}{\partial z^2}\,dz. \tag{5.77}$$

Here

$$S_1 = a\int_0^1 x\,\frac{\partial^2 x}{\partial z^2}\,dz, \tag{5.78}$$

according to Eq.(5.55). Integration by parts and inserting the boundary conditions finally yields the estimate

$$S_1 \leq -a(u_{2s}+\xi_2)x^2(t,0) - ax_s(0)\xi_2 x(t,0), \tag{5.79}$$

hence in terms of Eq.(5.56):

$$a_o = -\,au_{2s} < 0;\; a_1 = -a;\; a_2 = 0; \qquad a_3 = -\,\frac{a}{2}\,x_s(0);\; a_4 = 0. \tag{5.80}$$

The matrix $\underline{Q}$ is obtained as

$$\underline{Q}(k,u_2) = \begin{bmatrix} \underline{\tilde{A}}^T\underline{P}+\underline{P}\,\underline{\tilde{A}} & a_3\underline{e}_2-k\underline{P}\,\underline{b}^* \\ a_3\underline{e}_2^T-k\underline{b}^{*T}\underline{P} & a_o+a_1u_2 \end{bmatrix} . \tag{5.81}$$

The question if $\dot{V} \equiv 0$ may occur along some perturbed motion can easily be answered. $\underline{\eta}(t,0) \equiv 0$ implies $x(t,0) \equiv 0$ and $\underline{\xi} \equiv 0$, hence $u_2(t) \equiv 0$. Therefore, solutions to the following set of equations have to be investigated:

$$\frac{\partial x}{\partial t} = a\,\frac{\partial^2 x}{\partial z^2}\,,\quad 0 \leq z \leq 1;$$

$$\left.\frac{\partial x}{\partial z}\right|_{z=0} = \left.\frac{\partial x}{\partial z}\right|_{z=1} = 0 \quad \textit{and} \quad x(t,0) = 0;$$

$$x(0,z) = x_o(z).$$

The general solution of the heat equation with boundary conditions of the second kind at z = 0 and ż = 1 is [18]

$$x(t,z) = \sum_{i=1}^{\infty} x_i^*(0) e^{\lambda_i t} \varphi_i(z),$$

where $x_i^*(0)$, i=1,2,..., are Fourier-coefficients of $x_o(z)$; $\varphi_1(z)=1$ and $\varphi_i(z)=\sqrt{2}\cos(i-1)\pi z$, i=2,3,..., are the eigenfunctions of the spatial operator and $\lambda_i = -a(i-1)^2\pi^2$, i=1,2,..., are the corresponding eigenvalues.

Now the additional condition $x(t,0) \equiv 0$ requires

$$x_1^*(0) + \sqrt{2} \sum_{i=2}^{\infty} x_i^*(0) e^{\lambda_i t} \equiv 0,$$

which, because of the linear independent terms, can be satisfied only if $x_i^*(0) = 0, i = 1, 2, 3, \ldots$, hence $x_o(z) \equiv 0$. This assures that $\dot{V} \not\equiv 0$ along any perturbed motion.

From Eq.(5.63) $\underline{P}$ is obtained as

$$\underline{P} = \frac{\gamma}{6} \begin{bmatrix} 11 & -3 \\ -3 & 2 \end{bmatrix},$$

hence

$$-\underline{Q}(k,u_2) = \begin{bmatrix} \gamma & 0 & \frac{11}{6}\gamma k \\ 0 & \gamma & -\frac{1}{2}\gamma k - a_3 \\ \frac{11}{6}\gamma k & -\frac{1}{2}\gamma k - a_3 & a(u_{2s}+u_2) \end{bmatrix}.$$

A simple calculation shows that

$$\gamma > \frac{a}{4} \frac{x_s^2(0)}{u_{2s}}$$

assures positive definiteness of $-\underline{Q}(0,0)$.

Evaluation of Eq.(5.67) now yields the parabola

$$\left(k + \frac{9a_3}{65\gamma}\right)^2 - \frac{18a}{65\gamma}\cdot(u_{2s} + u_2) + \frac{1089a_3^2}{(65\gamma)^2} = 0.$$

The following numerical example of the plant has been considered for the further computation:

$$a = 5;\ u_{2s} = 5;\ u_{3s} = 1,$$

hence

$$x_s(z) = z + 0{,}2.$$

Perturbation $u_2(t)$ may be bounded by

$$|u_2(t)| \leqq 1.$$

The parabola in the k, u_2-plane is of the type as outlined in Fig.5.5. The parameter $\gamma = 0{,}0252$ has been determined such as to make k_1 in Fig.5.5 as negative as possible.

Fig.5.10 outlines the parabola in the k, u_2-plane obtained for this numerical example. $k = -\ 7{,}9$ has been specified again from the requirement that the strip $|u_2| \leqq 1$ belong to the domain where $\dot{V} \leqq 0$.

The largest assurable domain of asymptotic stability in the ξ_1, ξ_2, $||x||$ - space due to Eq.(5.71) is obtained for $\beta =$ $= 0{,}002496$ and is outlined in Fig.5.11.

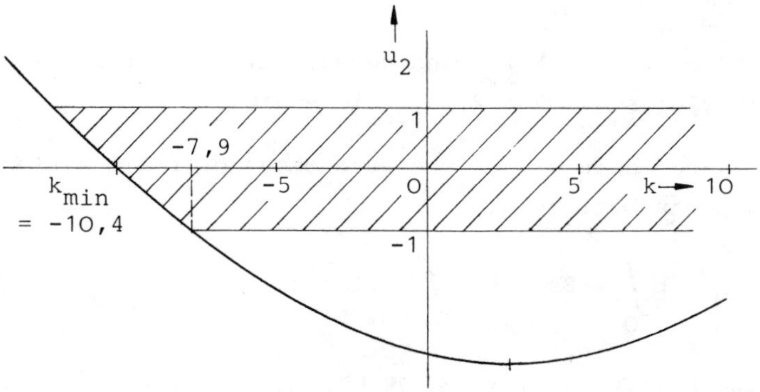

Fig.5.10: Domain of negative definite $\dot{V}_{max}$ in the k, u_2-plane; specification of $k = -\ 7{,}9$ from bounded u_2

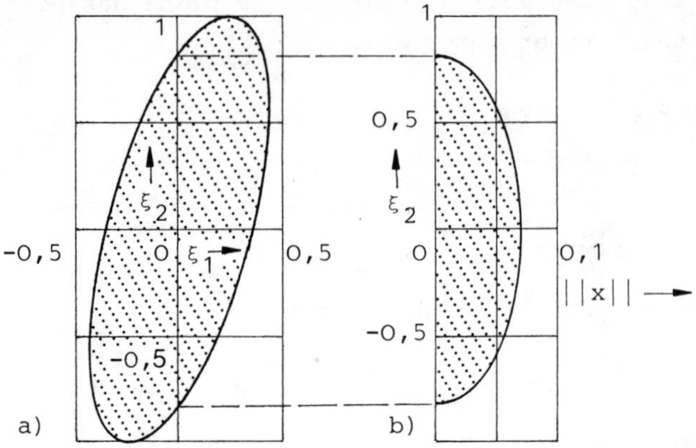

Fig.5.11: Domain of asymptotic stability in the ξ_1, ξ_2, $||x||$ - space

5.2.2 State feedback control

In this section a weakened version of the bang-bang control law from section 5.1 will be discussed in so far as linear lumped parameter feedback of the quadratic functional of the total state,

$$y(t) = \int_0^1 x \cdot Bx dz + \int_0^1 C(z) \cdot x dz,$$

according to Eq.(5.47) is considered. The preliminaries and assumptions of section 5.2 may hold again.

Now the state equations of the autonomous over-all system in terms of perturbations are

$$\dot{\underline{\xi}} = \underline{\tilde{A}}\,\underline{\xi} - \underline{\tilde{b}} \int_0^1 x \cdot Bx dz - \underline{\tilde{b}} \int_0^1 C(z) \cdot x dz, \tag{5.82}$$

$$\frac{\partial x}{\partial t} = Ax + \xi_m \cdot Bx + C(z) \cdot \xi_m, \quad \xi_m = u, \tag{5.83}$$

$$Rx = 0, \tag{5.84}$$

$$m \leq u(t) \leq M, \tag{5.85}$$

$$x(0,z) = x_o(z),\ \underline{\xi}(0) = \underline{\xi}_o. \tag{5.86}$$

The Ljapunow-functional

$$V = \underline{\xi}^T \underline{P}\, \underline{\xi} + \frac{1}{2}\int_0^1 x^2 dz \tag{5.87}$$

will be chosen again, hence

$$\begin{aligned}\dot{V} &= \underline{\dot{\xi}}^T \underline{P}\, \underline{\xi} + \underline{\xi}^T \underline{P}\, \underline{\dot{\xi}} + \int_0^1 x \frac{\partial x}{\partial t}\, dz = \\ &= \underline{\xi}^T (\underline{\tilde{A}}^T \underline{P} + \underline{P}\, \underline{\tilde{A}})\underline{\xi} + \\ &+ \int_0^1 x\cdot Axdz + (\underline{e}_m - 2\underline{P}\, \underline{\tilde{b}})^T \underline{\xi} \int_0^1 x\cdot Bxdz + \\ &+ (\underline{e}_m - 2\underline{P}\, \underline{\tilde{b}})^T \underline{\xi} \int_0^1 C(z)\cdot xdz.\end{aligned} \tag{5.88}$$

Comparison of the sum S_2 of integral terms in Eq.(5.88) with S_1, Eq.(5.55), shows that merely ξ_m has been replaced by $\underline{q}^T \underline{\xi}$, where

$$\underline{q} = \underline{e}_m - 2\underline{P}\, \underline{\tilde{b}}. \tag{5.89}$$

Therefore the same estimate as in Eq.(5.56) can be used. Moreover, a quadratic form in $\underline{\eta}(t,z_1)$, with any $z_1 \in [0,1]$ instead of z_M is appropriate now, as no specific single point of measurement is considered. This makes

$$\dot{V} \leqq \underline{\eta}^T(t,z_1)\underline{Q}\, \underline{\eta}(t,z_1), \tag{5.90}$$

where

$$\underline{Q} = \begin{bmatrix} \underline{\tilde{A}}^T \underline{P} + \underline{P}\, \underline{\tilde{A}} + a_2 \underline{q}\, \underline{q}^T & (a_3 + a_4 \underline{q}^T \underline{\xi})\underline{q} \\ (a_3 + a_4 \underline{q}^T \underline{\xi})\underline{q}^T & a_o + a_1 \underline{q}^T \underline{\xi} \end{bmatrix}. \tag{5.91}$$

If now again the lumped system has yet undetermined gain k, then

$$\underline{\tilde{b}} = k\underline{b}^*$$

with specified $\underline{b}^*$ and $\underline{\tilde{A}}$. This makes

$$\underline{q} = \underline{e}_m - 2k\underline{P}\,\underline{b}^* = \underline{q}(k), \tag{5.92}$$

hence

$$\underline{Q} = \underline{Q}(k,\underline{\xi}). \tag{5.93}$$

Different from Eq.(5.58), the domain of negative definiteness of $\underline{Q}$ is now a domain in the (m+1)-dimensional space spanned by k and $\underline{\xi}$. But all considerations from section 5.2.1 apply accordingly.

The special case where $a_3 = a_4 = 0$ becomes even simpler, as now $\underline{Q}$ takes the special form:

$$\underline{Q}(k,\underline{\xi}) = \begin{bmatrix} \underline{\tilde{A}}^T\underline{P}+\underline{P}\,\underline{\tilde{A}} + a_2\underline{q}\,\underline{q}^T & \underline{0} \\ \underline{0}^T & a_o + a_1\underline{q}^T\underline{\xi} \end{bmatrix} . \tag{5.94}$$

Hence Sylvester's criterion falls into the two requirements that

1. $-\underline{\tilde{A}}^T\underline{P}-\underline{P}\,\underline{\tilde{A}} - a_2\underline{q}\,\underline{q}^T$ satisfy Sylvester's inequalities and (5.95)
2. $a_o + a_1\underline{q}^T\underline{\xi} < 0.$ (5.96)

Example 5.5:

The situation may be illustrated using example 5.3 with the only difference that now, in view of example 5.1 and Eq.(5.47),

$$y(t) = \frac{1}{2}\,x^2(t,1) + \int_0^1 x_s'(z)x(t,z)dz \tag{5.97}$$

is fed back instead of $y(t) = x(t,1)$.

To start with requirement (5.96) we have now

$$-\frac{1}{2} - \frac{1}{2}\cdot\left[-\frac{11}{3}\gamma k\xi_1 + (1+\gamma k)\xi_2\right] < 0,$$

hence

$$\xi_2 > -\frac{1}{1+\gamma k} + \frac{\frac{11}{3}\gamma k}{1+\gamma k}\,\xi_1, \tag{5.98}$$

which is a straight line in the ξ_1, ξ_2-plane with parameter γk.

The requirement of

$$- \underline{\tilde{A}}^T \underline{P} - \underline{P}\,\underline{\tilde{A}} - a_2 \underline{q}\,\underline{q}^T = \gamma \underline{I} - a_2 \underline{q}\,\underline{q}^T$$

to satisfy Sylvester's inequalities provides the following numerical conditions on γ and k:

$$\gamma k^2 < 0{,}69 \qquad (5.99)$$

and

$$(\gamma k)^2 + 0{,}1385\gamma k - 0{,}642(\gamma - 0{,}108) < 0 \qquad (5.100)$$

$(\gamma > a_2 = 0{,}108)$.

Conditions (5.99) and (5.100) specify a domain in the γ, k-plane graphed in Fig.5.12. It can be seen that for $\gamma = 0{,}165$ the admissible k reaches its maximum:

$$k_{max} = 1{,}63.$$

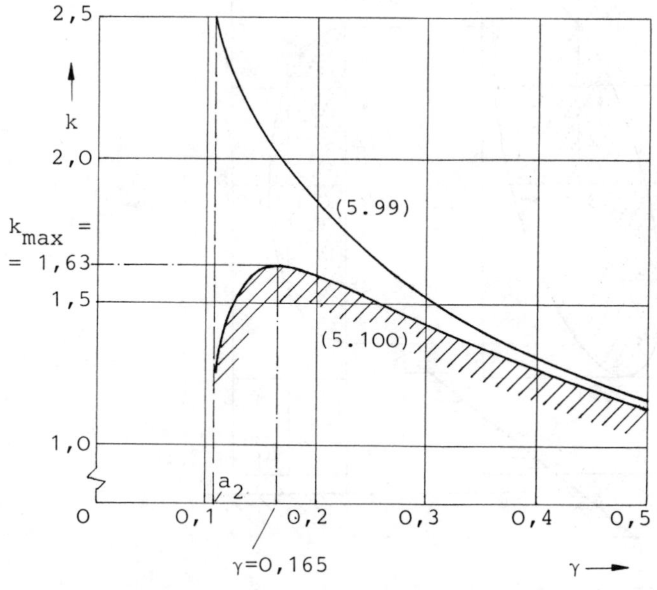

Fig.5.12: Inequalities (5.99) and (5.100) in the γ, k-plane

Therefore $\gamma = 0,165$ and $k = 1,6$ may be assumed satisfying condition (5.95) on the one hand and specifying, with regard to Eq.(5.98), the domain U of negative definiteness of $\underline{Q}$ on the other hand (Fig.5.13):

$$U = \{\underline{\xi}(t) \mid \xi_2 > -0,79 + 0,765\xi_1\}.$$

Then in terms of the total state of the system, $\underline{\eta}(t,z)$, we have

$$\dot{V} \leq 0 \text{ for each } \underline{\eta}(t,z) \in \Theta := \{\underline{\eta}(t,z) \mid \underline{\xi}(t) \in U\}. \quad (5.101)$$

From Fig.5.13 it can be seen that in the ξ_1, ξ_2, $\|x\|$ -space the domain Λ, determined as in example 5.3 (Fig.5.7), lies entirely in the interior of the domain, defined by $\underline{\xi} \in U$. The parameter β could even be enlarged in this case, but then the bounds on the control, $|u_1(t)| \leq 0,5$, could no longer be guaranteed.

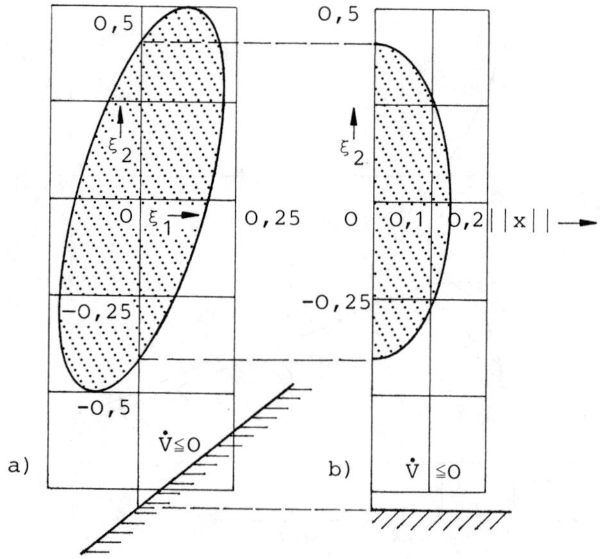

Fig.5.13: Domains of negative semidefinite $\dot{V}$ and of asymptotic stability in the ξ_1, ξ_2, $\|x\|$-space

5.3 Summary

Ljapunow's direct method has been applied to the closed-loop control of a class of bilinear distributed parameter plants. In section 5.1 very efficient control laws have been obtained due to the absence of any additional dynamical subsystems. In section 5.2 a linear lumped dynamical subsystem has been assumed. By valuing the results obtained in the latter case from a critical standpoint, it can be summarized that on the one hand analytical stability conditions are obtained directly from the mathematical model without the need of approximations. But on the other hand, because of the merely sufficient nature of these conditions, nothing can be said about their sharpness. Especially, it can be observed from example 5.3 that

a) the actual upper bound on the controller gain, with regard to stability, is wider than the assured one,

b) the actual domain of asymptotic stability in the state space is larger than the assured one,

c) the prescribed bounds on the control variable are *assured*, but not by far *exhausted*.

An expedient may be based on the following idea.

Provide two alternative control laws and a suitable strategy for the switching between them, as outlined in Fig.5.14.

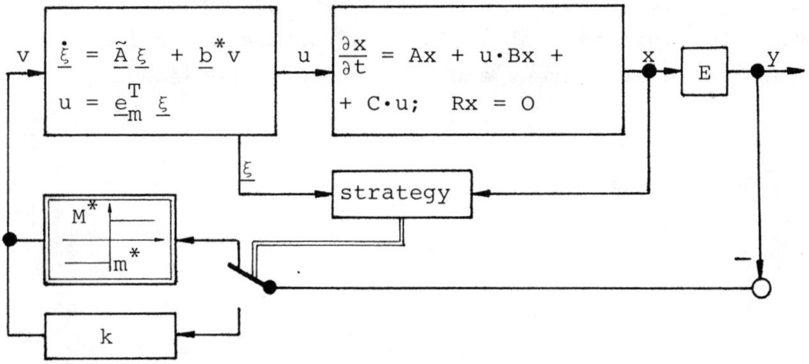

Fig.5.14: Variable structure control of the mixed lumped and distributed system

The output y(t) may be specified according to Eq.(5.46) or (5.47). The parameter k is assumed to be determined via methods of sections 5.2.1 or 5.2.2. Therefore, control law

$$v(t) = -ky(t) \tag{5.102}$$

will make the over-all system asymptotically stable with specifyable domain of asymptotic stability in the space spanned by $\underline{\xi}$ and $||x||$.

Application of the alternative control law

$$v(t) = \frac{M^*+m^*}{2} + \frac{M^*-m^*}{2} \operatorname{sgn}[-y(t)], \tag{5.103}$$

(with yet unspecified m^* and M^*) during some *finite interval of time* and then passing over to the weak control law (5.102) is expected to improve performance of the controlled plant considerably.

Following this concept, three questions must be given an answer:

1. How to specify bounds m^* and M^* in view of bounded u(t),

2. How to specify a switching strategy and

3. How to assure an enlarged domain of asymptotic stability.

It should be noticed that, different from Fig.4.4, the structure sketched in Fig.5.14 is *not* piecewise linear. A design method for this type of variable structure control systems is being tested at present and will be published soon.

6. Opitimal control via Butkovskiy's maximum principle

6.1 Butkovskiy's fundamental integral equations

A. G. Butkovskiy [17] has formulated a very comprehensive maximum principle which provides necessary conditions for the optimal control of a broad class of linear and nonlinear lumped and distributed parameter systems. Butkovskiy's theory is basing on state equations in integral form, and therefore it is an appropriate tool especially in those distributed parameter problems where the Green's function of some linear subsystem is available. Since bilinear systems represent a special class of nonlinear systems, Butkovskiy's theory can be expected to give useful hints for the design of optimal and near-optimal bilinear distributed parameter control systems. Some interest will also be devoted to systems with mixed lumped and distributed nature (section 6.3). In general, the application of Butkovskiy's maximum principle will not provide explicit solutions to the optimal open- or closed-loop control problem. In most situations a set of coupled nonlinear integral equations is obtained which the optimal control and the corresponding optimal state must satisfy. Therefore, effective numerical methods are of great importance.

The fundamental statement of Butkovskiy's maximum principle will briefly be reported as follows:

Let the state of a dynamical system, lumped or distributed, be characterized by a vector $\underline{x}(P)$, where P is a point in a certain region D of an Euclidean space. Let $\underline{u}(P)$, $P \in D$, be a control vector, taking values from some admissible set S, and let the motion of the system, subject to control $\underline{u}(P)$, be represented by a nonlinear integral equation of the type

$$\underline{x}(P) = \int_D \underline{f}[P,Q,\underline{x}(Q),\underline{u}(Q)]dQ, \tag{6.1}$$

with specified vector function $\underline{f}$. The components of $\underline{f}$, f_i, are supposed to have continuous partial derivatives $\partial f_i/\partial x_j$ which constitute the square matrix

$$\frac{\partial \underline{f}}{\partial \underline{x}} = \left(\frac{\partial f_i}{\partial x_j}\right), \qquad (6.2)$$

where i denotes the row number and j the column number.

Then the optimal control problem consists in finding an admissible control $\underline{u}(P)$, $P \in D$, $\underline{u} \in S$, such as to minimize a preassigned cost functional of the form

$$J = \int_D v[P,\underline{x}(P), \underline{u}(P)]dP. \qquad (6.3)$$

The scalar function v is supposed to have continuous partial derivatives $\partial v/\partial x_j$ which constitute the row vector

$$\frac{\partial v}{\partial \underline{x}} = \left(\frac{\partial v}{\partial x_j}\right) . \qquad (6.4)$$

Then for the optimality of the control $\underline{u}(P)$, $P \in D$, $\underline{u} \in S$, it is necessary that for almost all fixed values of $P \in D$ the function

$$\Pi(P,\underline{u}) = - v[P,\underline{x}(P),\underline{u}] + \int_D \underline{\psi}^T(Q)\underline{f}[Q,P,\underline{x}(P),\underline{u}]dQ \qquad (6.5)$$

attains a maximum with respect to $\underline{u} \in S$.

Thereby the adjoint row vector, $\underline{\psi}^T(Q)$, satisfies the integral equation

$$\underline{\psi}^T(Q) + \partial v[Q,\underline{x}(Q),\underline{u}(Q)]/\partial \underline{x} =$$
$$= \int_D \underline{\psi}^T(P)\partial \underline{f}[P,Q,\underline{x}(Q),\underline{u}(Q)]/\partial \underline{x} dP. \qquad (6.6)$$

A certain difficulty in the application of this comprehensive theory arises from the requirement of all integrations to be made over the same domain D. Especially when treating distributed parameter systems it is therefore sometimes necessary to introduce fictive new state variables. Details can be seen from the application to bilinear distributed parameter systems in the following sections.

6.2 Optimal control of a class of bilinear distributed plants

6.2.1 Statement of the problem

The considerations of this section are based on the integral representation of bilinear distributed parameter systems as described in section 2.4. As pointed out in section 2.4, integro-differential equations will generally result from the Green's function method. Only the class of systems which are described by pure integral equations (example 2.6) or which can be transformed into pure integral equations via integration by parts (example 2.7) will be considered here. This is an essential assumption, since the integral state equation (6.1) in Butkovskiy's maximum principle does not admit $\underline{f}$ to depend on derivatives of $\underline{x}$.

A very general discussion of optimal bilinear distributed parameter processes would exceed the scope of this contribution. Therefore the following exemplary problem will be considered.

Let the integral state equation of the distributed plant be given by

$$x(t,z) = \int_{\tau=0}^{t} g_1(t-\tau,z)x(\tau,0)u(\tau)d\tau + \int_{\zeta=0}^{1} g(t,z,\zeta)x_o(\zeta)d\zeta , \quad 0 \leq z \leq 1, \tag{6.7}$$

with scalar state $x(t,z)$, scalar multiplicative control variable $u(t)$, initial state $x_o(z)$ and specified functions $g(t,z,\zeta)$ and $g_1(t-\tau,z)$.

Consider as a practical example the cooling of a massive body from the boundary, employing the flow rate of the cooling medium as a multiplicative control action. The problem has been modelled in example 2.6, Eq.(2.43). If u_1 and u_3 are assumed to vanish in Eq.(2.43), then in terms of Eq.(6.7)

$$g_1(t-\tau,z) = - a \cdot g(t-\tau,z,0), \tag{6.8}$$

where $g(t-\tau,z,\zeta)$ is the Green's function of the linear part of the problem.

The admissible set, S, of control u(t) may be given by

$$S = \{u(t)\,|\,m \leq u(t) \leq M\}, \tag{6.9}$$

not necessarily containing the origin.

Before specifying a criterion of performance, some fundamental remarks should be made with regard to reachability (section 3.3). In many situations the objective of optimal control consists in attaining a specified state $x(T,z) = w(z)$ in finite time T. Therefore, depending on whether or not $w(z) \in X_T(x_o)$ the following problems of optimal control arise:

a) If $w(z) \in X_T(x_o)$ with specified T, then $x(T,z) = w(z)$ is a boundary condition at $t = T$, and a criterion of performance may be

$$J_o = \frac{1}{2} \int_{t=0}^{T} u^2(t)\,dt, \tag{6.10}$$

in order to minimize the control energy.

b) If $w(z) \in X_T(x_o)$ for some T, then the additional problem may be to find the least value T_{min}, in which $x(T,z) = w(z)$ is attainable. This is known as the time-optimal problem.

c) If $w(z) \notin X_T(x_o)$ with specified T, then the optimal control problem must be posed in a different manner. Now the criterion of performance

$$J_1 = \frac{1}{2} \int_{z=0}^{1} [w(z) - x(T,z)]^2 dz \tag{6.11}$$

is a measure for the mean deviation of the actual final state $x(T,z)$ from the desired distribution $w(z)$. Eq. (6.11) may be augmented by an energy term to obtain

$$J_2 = \frac{1}{2} \int_{z=0}^{1} [w(z) - x(T,z)]^2 dz + \frac{\gamma}{2} \int_{t=0}^{T} u^2(t)\,dt, \tag{6.12}$$

where γ is a positive weighting factor.

Situation c) seems to be of greatest importance especially for homogeneous-in-the-state bilinear systems with restricted reachability properties (section 3.3.2). As the underlying state equation (6.7) is of this type, only performance criterions (6.11) and (6.12) will be discussed in the following, hence $w(z) \notin X_T(x_o)$ is assumed.

By taking the process of multiplicative boundary cooling as an example, a desirable terminal distribution may be

$$w(z) \equiv \text{const.}, \quad 0 \leq z \leq 1. \tag{6.13}$$

But from boundary condition (2.11),

$$\left.\frac{\partial x(t,z)}{\partial z}\right|_{z=0} = u_2(t)x(t,0),$$

it can directly be concluded, that $w(z) \equiv \text{const.} (\neq 0)$ is not reachable from any $x_o(z)$ with $(0<)m_2 \leq u_2(t) \leq M_2$. This makes J according to Eq.(6.11) or (6.12) an adequate criterion of performance.

6.2.2 The optimal terminal state problem

Now the J_1-optimal problem, Eq.(6.11), will be considered. Here D can be made to be the time-interval [0,T] by means of the following procedure:

Write Eq.(6.7) for z = 0 and define the time-dependent state variable

$$x_1(t) = x(t,0) \tag{6.14}$$

to obtain

$$x_1(t) = \int_{\tau=0}^{t} g_1(t-\tau,0)x_1(\tau)u(\tau)d\tau + \tilde{x}_o(t), \tag{6.15}$$

where

$$\tilde{x}_o(t) = \int_{\zeta=0}^{1} g(t,0,\zeta)x_o(\zeta)d\zeta \tag{6.16}$$

is the motion of $x_1(t)$ if $u \equiv 0$ were applied.

Next, write the u-dependent terms of J_1 as an integral of time by inserting the detailed expression for $x(T,z)$ into J_1 as follows:

$$J_1 = \frac{1}{2}\int_{z=0}^{1}[w(z) - \int_{\zeta=0}^{1} g(T,z,\zeta)x_o(\zeta)d\zeta]^2 dz +$$

$$+ \frac{1}{2}\int_{z=0}^{1}[\int_{\tau=0}^{T} g_1(T-\tau,z)x_1(\tau)u(\tau)d\tau]^2 dz -$$

$$- \int_{z=0}^{1}[w(z) - \int_{\zeta=0}^{1} g(T,z,\zeta)x_o(\zeta)d\zeta] \cdot$$

$$\cdot \int_{\tau=0}^{T} g_1(T-\tau,z)x_1(\tau)u(\tau)d\tau \cdot dz. \qquad (6.17)$$

The first right hand integral in Eq.(6.17) can be omitted, as it does not depend on the state or on the control. The second integral can be rearranged by writing

$$[\int_{\tau=0}^{T}(.)d\tau]^2 = \int_{t=0}^{T}\int_{\tau=0}^{T}(.)(.)d\tau dt.$$

Finally, by changing the order of integrations, one obtains an expression of the form

$$J_1 = \frac{1}{2}\int_{t=0}^{T}[x_2(t) - 2R(t)]x_1(t)u(t)dt, \qquad (6.18)$$

with fictive state variable $x_2(t)$ satisfying

$$x_2(t) = \int_{\tau=0}^{T} K(t,\tau)x_1(\tau)u(\tau)d\tau. \qquad (6.19)$$

The following abbreviations have been used:

$$K(t,\tau) = \int_{z=0}^{1} g_1(T-\tau,z)g_1(T-t,z)dz, \qquad (6.20)$$

$$R(t) = \int_{z=0}^{1} \tilde{w}(z) g_1(T-t,z) dz, \tag{6.21}$$

$$\tilde{w}(z) = w(z) - \int_{\zeta=0}^{1} g(T,z,\zeta) x_o(\zeta) d\zeta. \tag{6.22}$$

Now the whole problem of optimization can be rewritten in the form required in section 6.1:

The state vector $\underline{x}(t) = [x_1(t), x_2(t)]^T$ satisfies

$$\underline{x}(t) = \int_{\tau=0}^{T} \underline{f}[t,\tau,\underline{x}(\tau),u(\tau)] d\tau, \tag{6.23}$$

with

$$f_1 = g_1(t-\tau,0) x_1(\tau) u(\tau) \sigma(t-\tau) + \tilde{x}_o(t) \delta(\tau) \tag{6.24}$$

and

$$f_2 = K(t,\tau) x_1(\tau) u(\tau), \tag{6.25}$$

$\sigma(t)$ and $\delta(t)$ denoting the unit step function and the Dirac function, respectively.

It should be noted that this fictive dynamical system is non-causal as $K(t,\tau) \not\equiv 0$ for $\tau > t$. But this is admissible in Butkovskiy's theory.

The problem consists in finding $u(t)$, $m \leqq u(t) \leqq M$, such that J_1, Eq.(6.18), assumes its smallest possible value.

In terms of Eq.(6.3) now

$$v[t,\underline{x}(t),u(t)] = \frac{1}{2}[x_2(t) - 2R(t)] x_1(t) u(t), \tag{6.26}$$

hence

$$\frac{\partial v}{\partial \underline{x}} = \frac{1}{2} u(t) \cdot [x_2(t) - 2R(t), x_1(t)]. \tag{6.27}$$

Moreover

$$\frac{\partial \underline{f}}{\partial \underline{x}} = u(\tau)\cdot \begin{bmatrix} g_1(t-\tau)\sigma(t-\tau) & 0 \\ K(t,\tau) & 0 \end{bmatrix}. \qquad (6.28)$$

By means of these preparations the following explicit expression is obtained from Eq.(6.6) for the adjoint variable $\Psi_2(\tau)$:

$$\Psi_2(\tau) = -\frac{1}{2} x_1(\tau)u(\tau), \qquad (6.29)$$

whereas $\Psi_1(\tau)$ satisfies a Volterra integral equation:

$$\Psi_1(\tau) - \int_{t=\tau}^{T} \Psi_1(t) g_1(t-\tau,0)u(\tau)dt = [R(\tau)-x_2(\tau)]u(\tau). \qquad (6.30)$$

Now the Π-function according to Eq.(6.5) will be considered:

$$\Pi(t,u) = -\frac{1}{2}[x_2(t)-2R(t)]x_1(t)u +$$

$$+ \sum_{i=1}^{2} \int_{\tau=0}^{T} \Psi_i(\tau) f_i[\tau,t,x_1(t),u]d\tau \stackrel{!}{=} \max_{m \leqq u \leqq M}. \qquad (6.31)$$

Only the u-dependent terms need to be considered which finally provides the necessary condition

$$u\cdot x_1(t)\cdot[R(t)-x_2(t)+\int_{\tau=t}^{T} \Psi_1(\tau)g_1(\tau-t,0)d\tau] \stackrel{!}{=} \max_{m \leqq u \leqq M}. \qquad (6.32)$$

This requirement will be satisfied by the control

$$u(t) = \frac{1}{2}(M+m) + \frac{1}{2}(M-m)\operatorname{sgn}[x_1(t)]\cdot$$

$$\cdot \operatorname{sgn}[R(t)-x_2(t) + \int_{\tau=t}^{T} \Psi_1(\tau)g_1(\tau-t,0)d\tau]. \qquad (6.33)$$

This is the optimal control $u_{opt}(t)$, presumed that the singular case to be discussed in section 6.2.3 does not occur. Obviously, $u_{opt}(t)$ is of the bang-bang-type taking only values m and M.

If the interval [m,M] does not contain the origin and therefore u = 0 can be excluded, Eq.(6.33) can be simplified by noting that with the aid of Eq.(6.30) now

$$R(t) - x_2(t) + \int_{\tau=t}^{T} \Psi_1(\tau) g_1(\tau-t,0)\, d\tau = \frac{\Psi_1(t)}{u(t)} . \tag{6.34}$$

For the example where m, M>0, Eq.(6.33) now takes the special form

$$u(t) = \frac{1}{2}(M+m) + \frac{1}{2}(M-m)\,\mathrm{sgn}[x_1(t)\Psi_1(t)]. \tag{6.35}$$

Unfortunately, Eqs.(6.33) and (6.35) are far from being the *explicit* solution to the optimal control problem. The set of coupled nonlinear integral equations (6.23), (6.30) and (6.33) or (6.35), respectively, has to be solved simultaneously.

6.2.3 Singular control

From Eqs.(6.33) and (6.35) it can be concluded that the J_1-optimal control is not defined if during a finite interval of time $x_1(t) \equiv 0$ or $\Psi_1(t) \equiv 0$.

$x_1(t) \equiv 0$ during some interval $[t_1, t_2]$ requires, with regard to the time-invariant bilinear system (6.7), that

$$\int_{\zeta=0}^{1} g(t-t_1,0,\zeta)\, x(t_1,\zeta)\, d\zeta = 0 \text{ for all } t\in[t_1,t_2]. \tag{6.36}$$

Now let

$$\{\varphi_k(\zeta)\}, \quad \zeta\in[0,1], \quad k = 1, 2, \ldots, \tag{6.37}$$

be an arbitrary complete orthonormal set and let $g(t,0,\zeta)$ be expanded in the series

$$g(t,0,\zeta) = \sum_{k=1}^{\infty} g_k(t)\, \varphi_k(\zeta), \tag{6.38}$$

with linear independent $g_k(t)$, k = 1, 2, ... Then Eq.(6.36) requires $x(t_1,\zeta)$ to be orthogonal to all $\varphi_k(\zeta)$, hence

$$x(t_1,z) \equiv 0, \quad z\in[0,1],$$

and therefore

$$x(t,z) \equiv 0, \quad (t,z)\in[t_1,t_2]\times[0,1]. \tag{6.39}$$

This situation can be excluded in many applications by physical considerations.

Now $\Psi_1(t) \equiv 0$ during some interval $[t_1, t_2]$ will be examined. With regard to Eq.(6.30) this requires

$$R(t_2) - x_2(t_2) = -\int_{\tau=t_2}^{T} \Psi_1(\tau) g_1(\tau-t_2,0)\,d\tau \;:=\kappa=\text{const.}, \tag{6.40}$$

and also

$$-\int_{\tau=t}^{T} \Psi_1(\tau) g_1(\tau-t,0)\,d\tau = \kappa \text{ for all } t\in[t_1,t_2], \tag{6.41}$$

hence

$$R(t) - x_2(t) \equiv \kappa = \text{const.}, \quad t\in[t_1,t_2]. \tag{6.42}$$

By definitions of $R(t)$ and $x_2(t)$ this means

$$\int_{z=0}^{1} g_1(T-t,z)\cdot[w(z)-x(T,z)]\,dz \equiv \kappa, \; t\in[t_1,t_2]. \tag{6.43}$$

The same procedure as applied above (Eqs.(6.37) and (6.38)) now by using an expansion for $g_1(t,z)$,

$$g_1(t,z) = \sum_{k=1}^{\infty} g_{1k}(t)\,\varphi_k(z), \tag{6.44}$$

indicates that Eq.(6.43) can be satisfied only if

$$x(T,z) \equiv w(z), \tag{6.45}$$

unless one of the $g_{1k}(t)$ is a constant, say $g_{1j}(t) \equiv g_{1j}^{*}$, in which case Eq.(6.43) requires

$$\int_{z=0}^{1} [w(z) - x(T,z)]\,\varphi_k(z)\,dz = \frac{\kappa}{g_{1j}^{*}}\cdot\delta_{jk}, \; k=1, 2, \ldots \tag{6.46}$$

and therefore

$$x(T,z) \equiv w(z) - \frac{\kappa}{g_{1j}^{*}} \varphi_j(z). \tag{6.47}$$

Eqs.(6.45) and (6.47) indicate that $\Psi_1(t) \equiv 0$ can occur only if by accident w(z) and T are pre-assigned such that $w(z) \in X_T(x_o)$ or $w(z) - \kappa\varphi_j(z)/g_{1j}^{*} \in X_T(x_o)$, respectively.

Indeed, if w(z) is reachable from $x_o(z)$ in time T, then J_1 takes its smallest possible value, zero, and therefore u(t) can no longer be determined from the J_1-optimal problem. This may conclude the discussion of singular control.

6.2.4 The mixed optimal terminal state and minimum energy problem

Minimization of performance criterion J_2 due to Eq.(6.12) needs only a slight modification of the considerations of section 6.2.2. The integral state equations (6.23) remain unchanged. Only J_1 according to Eq.(6.18) has to be augmented by an energy term, and therefore

$$J_2 = \frac{1}{2} \int_{t=0}^{T} \{[x_2(t) - 2R(t)]x_1(t)u(t) + \gamma u^2(t)\}dt. \tag{6.48}$$

But this different expression will not change $\partial v/\partial \underline{x}$ compared with Eq.(6.27). That is why Ψ_1 and Ψ_2 satisfy the same equations as for $\gamma = 0$.

Now the u-dependent part of the Π-function is

$$u \cdot x_1(t) \cdot [R(t) - x_2(t) + \int_{\tau=t}^{T} \Psi_1(\tau) g_1(\tau - t, 0) d\tau] - \gamma u^2.$$

In the case where $0 \notin [m,M]$ and therefore $u(t) \neq 0$, this expression can be simplified by means of Eq.(6.34) to obtain the following necessary condition of optimality:

$$u x_1(t) \frac{\Psi_1(t)}{u(t)} - \frac{\gamma}{2} u^2 \overset{!}{=} \max_{m \leq u \leq M}, \tag{6.49}$$

which, via quadratic completion with respect to the variable u, will be satisfied by

$$u = \begin{cases} m, \text{ if } \frac{1}{\gamma} x_1(t)\Psi_1(t)/u(t) \leq m; \\ \frac{1}{\gamma} x_1(t)\Psi_1(t)/u(t), \text{ if } m < \frac{1}{\gamma} x_1(t)\Psi_1(t)/u(t) < M; \\ M, \text{ if } \frac{1}{\gamma} x_1(t)\Psi_1(t)/u(t) \geq M. \end{cases} \qquad (6.50)$$

This implicit expression for the optimal control $u(t)$ can be rewritten, by means of some simple transformations, in the form $u(t) = F[x_1(t)\Psi_1(t)]$, with function F being specified by (Fig.6.1)

$$u(t) = \begin{cases} m, \text{ if } x_1(t)\Psi_1(t) \leq \gamma m^2; \\ \sqrt{x_1(t)\Psi_1(t)/\gamma}, \text{ if } \gamma m^2 < x_1(t)\Psi_1(t) < \gamma M^2; \\ M \text{ if } x_1(t)\Psi_1(t) \geq \gamma M^2. \end{cases} \qquad (6.51)$$

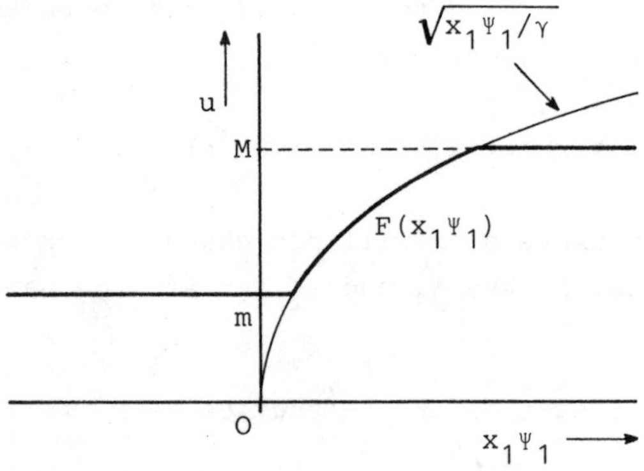

Fig.6.1: The nonlinear function $u = F(x_1\Psi_1)$ given by Eq.(6.51)

In the J_2-optimal problem this nonlinearity replaces the switching characteristic, Eq.(6.35), of the J_1-optimal problem. Of course, Eq.(6.51) tends to the switching characteristic of Eq.(6.35) as $\gamma \to 0$. On the other hand, $u(t) \equiv m$ as $\gamma \to \infty$. This is indeed what one would have expected intuitively.

As a consequence of continuity of the function F in Fig.6.1, u(t) will not be bang-bang-type and also the singular case will not occur in the J_2-optimal problem.

6.2.5 Computational aspects and numerical example

The set of coupled nonlinear integral equations (6.23), (6.30) along with Eqs.(6.35) or (6.51) in the J_1- or J_2-optimal problem have to be solved, respectively, in order to obtain the optimal open-loop control function u(t). Any computational method for solving nonlinear integral equations may be applied to the problem in hand. In the following, a basic iterative procedure will be proposed which may be employed regardless which method of approximation of the infinite-dimensional distributed parameter system will be used. After this preparation, eigenfunction series truncation will be applied as a method of approximation, to obtain the near-optimal bilinear boundary control of a distributed parameter cooling process.

The adjoint integral equation (6.30) represents a terminal-value problem and therefore plays a key role when constructing an iterative procedure for computing the optimal control. Eq.(6.30) will therefore be rewritten in the form

$$\Psi_{1,k+1}(t) = u_k(t)\cdot[R(t)-x_{2,k}(t)+\int_{\tau=t}^{T}\Psi_{1,k}(\tau)g_1(\tau-t,0)d\tau], \qquad (6.52)$$

where k denotes the number of iteration steps.

Fig.6.2 illustrates the basic signal flow diagram thus obtained for the iterative solution of the J_1-optimal problem. Again, Eq.(6.35) has simply to be replaced by Eq.(6.51) in the J_2-optimal problem.

As can be seen from Fig.6.2, there is only one interation loop. The feedback of $x_{1,k}(t)$ from Eq.(6.15) to Eq.(6.35) is not a serious problem, as Eqs.(6.15), (6.35) represent an *initial-value* problem which can be solved without iteration, by simple simulation. ε is a parameter of relaxation, which must be made small enough ($0 < \varepsilon < 1$) in order to obtain a convergent iteration process.

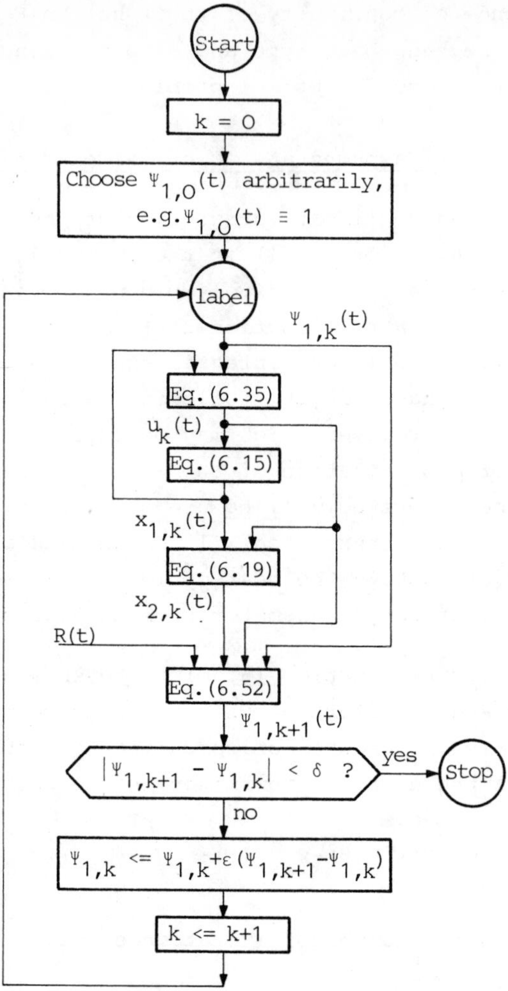

Fig.6.2: Basic signal flow diagram for iterative computation

For computing the problem in hand approximately, let the Green's function $g(t,z,\zeta)$ in Eq.(6.7) be representable by an infinite series,

$$g(t,z,\zeta) = \sum_{i=1}^{\infty} g_i(t)\,\varphi_i(z)\,\varphi_i(\zeta), \qquad (6.53)$$

where $\{\varphi_i(z)\}$ is a complete orthonormal set of eigenfunctions, and let $g_1(t-\tau,z)$ be expanded in the series

$$g_1(t-\tau,z) = \sum_{i=1}^{\infty} g_{1i}(t-\tau)\,\varphi_i(z). \tag{6.54}$$

In numerous control problems in the theory of distributed parameter systems the use of eigenfunction series is a proper way of rewriting the underlying equations in terms of Fourier coefficients depending on time t. Moreover, series truncation is an effective method to obtain approximate solutions which may be regarded to be near-optimal.

By means of Eqs.(6.53) and (6.54) and by using the truncated series

$$x(t,z) \approx \sum_{i=1}^{n} x_i^*(t)\,\varphi_i(z) \tag{6.55}$$

as an n-th order approximation of the state of the original system, one finally obtains from Eqs.(6.23), (6.30) and (6.35) the following set in case of J_1-near-optimality:

$$x_i^*(t) = \int_{\tau=0}^{t} g_{1i}(t-\tau)u(\tau)\sum_{j=1}^{n} x_j^*(\tau)\varphi_j(0)\,d\tau + x_i^*(0)g_i(t), \qquad i=1,\ \ldots,\ n, \tag{6.56}$$

$$\Psi_1(t) = u(t)\cdot\sum_{i=1}^{n}\Big\{\varphi_i(0)\cdot\int_{\tau=t}^{T} g_{1i}(\tau-t)\Psi_1(\tau)\,d\tau + [w_i^* - x_i^*(T)]g_{1i}(T-t)\Big\}, \tag{6.57}$$

$$u(t) = \frac{1}{2}(M+m) + \frac{1}{2}(M-m)\,\mathrm{sgn}[\Psi_1(t)\sum_{i=1}^{n} x_i^*(t)\,\varphi_i(0)]. \tag{6.58}$$

For the iterative solution of the set of Eqs.(6.56) - (6.58) with unknown functions $x_i^*(t)$, i=1, ..., n, $\Psi_1(t)$, u(t), a hybrid computer device can be used, very similar to that proposed by the author for the optimal control of linear distributed parameter systems [42]. The procedure of iteration makes use of the simple equality

$$\int_{\tau=t}^{T} (.)d\tau = \int_{\tau=0}^{T} (.)d\tau - \int_{\tau=0}^{t} (.)d\tau$$

which transforms the terminal-value problem, Eq.(6.57), into an initial-value problem, whose initial values are essentially influenced by $\Psi_1(t)$ from the foregoing iteration step. For more details see [42].

Example 6.1:

Consider the thermal process described in example 2.3, with additional assumptions that $u_1 \equiv 0$ and $u_3 \equiv 0$. Hence, Eqs. (2.10) - (2.13) for the parametric boundary cooling or heating process take a homogeneous-in-the-state form. The admissible set for the multiplicative control variable $u_2(t)$ may be specified by

$$S = \{u_2(t) \mid (0<)m_2 \leqq u_2(t) \leqq M_2\}.$$

Of course, the zero steady state of this process is asymptotically stable for arbitrary $u_2(t) \in S$ and for every $x_0(z)$, see section 3.2, example 3.6. Therefore, as $\|x\|$ is a monotonically decreasing function of time, the above system will have very restricted reachability properties as pointed out in section 3.3 (Remark 3).

It may be required that $x(T,z)$, with specified terminal time T, approach the nonreachable state

$$w(z) \equiv w_0, \quad z \in [0,1],$$

as closely as possible. As a concrete example, one may think of a reheating furnace for a rolling mill.

Hence the J_1-optimal problem according to Eq.(6.11) is a meaningful problem of optimal parametric control.

In detail, the Green's functions $g(t,z,\zeta)$ and $g_1(t-\tau,z)$ have already been specified in section 2.4, example 2.6. Therefore, the set of Eqs.(6.56) - (6.58) apply. The near-optimal control problem has been computed for the following numerical example:

$a = 1; \quad m = 0{,}25; \quad M = 1; \quad T = 0{,}45.$

$x_o(z) = 1, \quad 0 \leq z \leq 1,$

$w(z) = 0{,}7, \quad 0 \leq z \leq 1.$

Both, J_1-optimal and J_2-optimal control ($\gamma = 0{,}1$) have been considered. n = 2 has been specified, which will provide a rough idea of the optimal control and the corresponding optimal state. Small n will also be of interest from the practical point of view, as it keeps the number of switchings sufficiently small.

In Fig.6.3 the J_1-near-optimal functions $u(t)$, $x_1^*(t)$, $x_2^*(t)$ and $x(t,0) = x_1^*(t) + \sqrt{2}x_2^*(t)$ are graphed versus time. As can be seen from Fig. 6.4, the final state $x(T,z)$ compares favourably good with the desired state in view of restricted reachability properties.

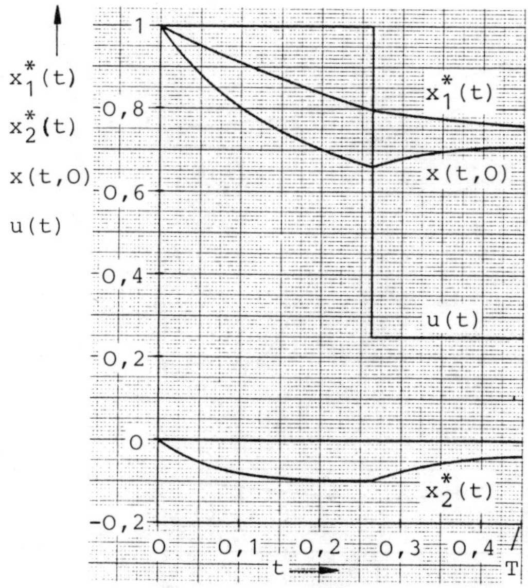

Fig.6.3: Numerical results of the J_1-near-optimal problem

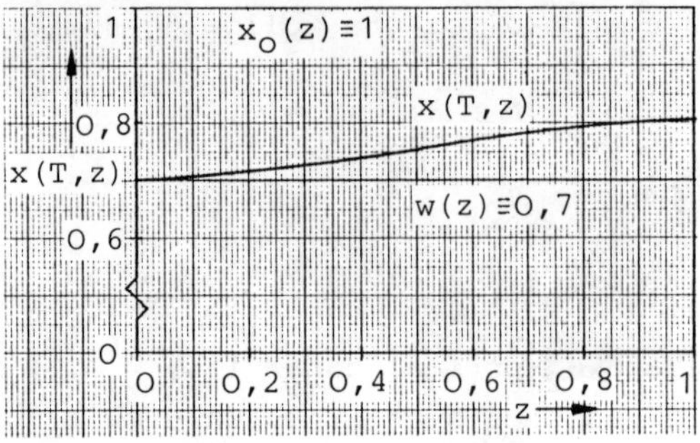

Fig.6.4: Terminal state reached by J_1-near-optimal control

In Fig.6.5 the J_2-near-optimal functions $u(t)$, $x_1^*(t)$, $x_2^*(t)$ and $x(t,0) = x_1^*(t) + \sqrt{2}x_2^*(t)$ are graphed versus time ($\gamma = 0,1$). Due to continuity of the function F in Fig.6.1, the near-optimal control u(t) is now continuous as well. Weighted control energy will of course, as an additional restriction, cause the deviation between x(T,z) and w(z) to be larger than in the J_1-near-optimal problem (Fig.6.6).

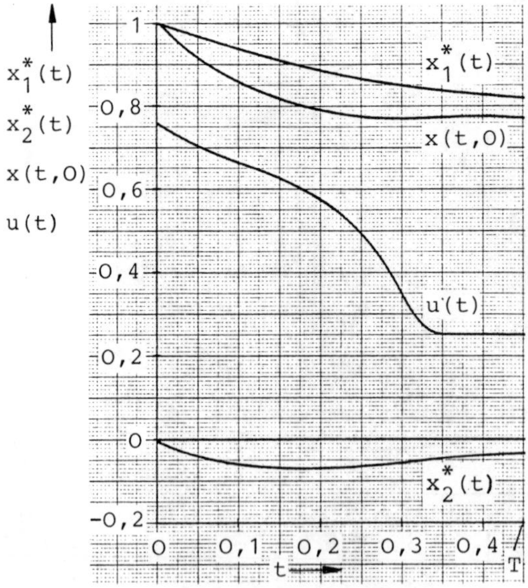

Fig.6.5: Numerical results of the J_2-near-optimal problem

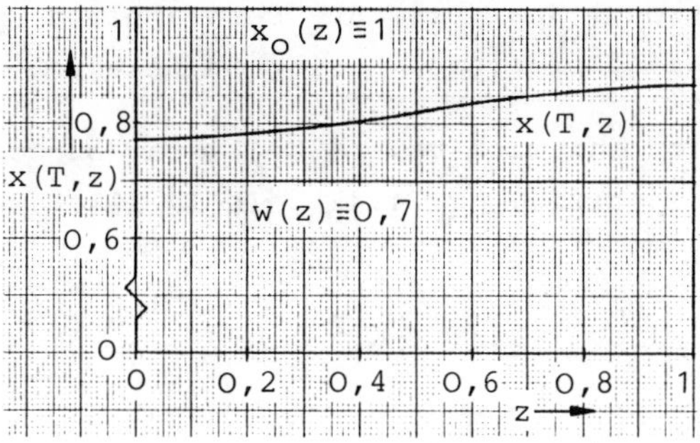

Fig.6.6: Terminal state reached by J_2-near-optimal control

6.3 Optimal variable structure closed-loop control

6.3.1 Statement of the problem

The mixed lumped and distributed control system described in section 4.2 will be considered now. The time-dependent gain matrix $\underline{K}(t)$ makes this variable structure control system a bilinear system. As prepared in section 4.2.2, the following integral equation will serve as a mathematical model of the over-all system:

$$\underline{y}(t) = \int_{\tau=0}^{t} \underline{\tilde{G}}(t-\tau)\underline{K}(\tau)\underline{y}(\tau)d\tau + \underline{y}_o(t), \tag{6.59}$$

all functions being specified in section 4.2. Especially, $\underline{K}(t)$ is assumed to be switchable between two fixed matrices, $\underline{K}_1$ and $\underline{K}_2$, according to Eqs.(4.35) and (4.36), which makes $\tilde{u}(t)$ act as a scalar control variable.

Additionally, $\underline{K}_1$ will be assumed to be pre-specified such that the *linear* system resulting from Eqs. (4.32) - (4.34) for $\underline{K}(t) \equiv \underline{K}_1$ is asymptotically stable, hence

$$\lim_{t\to\infty} \underline{\eta}(t,\underline{z}) = \underline{0} \quad \text{for } \underline{K}(t) \equiv \underline{K}_1. \tag{6.60}$$

This linear system property does of course not depend on the initial perturbation $\underline{\eta}_o(\underline{z})$. Frequency domain methods or Ljapunow's direct method may be applied to specify such a $\underline{K}_1$.

Methods of functional analysis have been applied by Schober [53] to this type of problems, basing on integral models.

By this assumption and with regard to linear output equation (4.30) or, more specific, Eq. (4.42), it is also assured that in Eq. (6.59)

$$\lim_{t\to\infty} \underline{y}(t) = \underline{O} \quad \text{for} \quad \underline{K}(t) \equiv \underline{K}_1 \tag{6.61}$$

for every initial perturbation.

The alternative gain matrix $\underline{K}_2$ may be specified arbitrarily, very different from $\underline{K}_1$, even such that $\underline{K}(t) \equiv \underline{K}_2$ would make an unstable over-all system.

Then the following problem of optimal control will be significant:

Determine a switching strategy for $\tilde{u}(t)$, $0 \leq t \leq T$, such that the performance criterion

$$J = \frac{1}{2} \int_{t=0}^{T} \underline{y}^T(t) \underline{P} \, \underline{y}(t) dt \tag{6.62}$$

takes its smallest possible value, where $\underline{P}$ is a positive definite matrix, and T is a specified terminal time.

The time interval [O,T] may be the time of putting into operation a large plant. Due to minimization of J, the vector $\underline{y}(T)$ will be in some neighbourhood of $\underline{O}$ and therefore $\underline{K}(t) \equiv \underline{K}_1$ may be employed for $t > T$.

6.3.2 Application of Butkovskiy's maximum principle

In terms of section 6.1 we have now

$$\underline{x}(P) \mathrel{\hat{=}} \underline{y}(t), \tag{6.63}$$

$$\underline{f}[t,\tau,\underline{y}(\tau),\tilde{u}(\tau)] = \underline{\tilde{G}}(t-\tau)[\tfrac{1}{2}(\underline{K}_1+\underline{K}_2) + $$

$$+ \tfrac{1}{2}(\underline{K}_1-\underline{K}_2)\tilde{u}(\tau)]\underline{y}(\tau)\sigma(t-\tau) + \underline{y}_o(t)\delta(\tau), \tag{6.64}$$

$$D = [O,T], \tag{6.65}$$

$$v = \frac{1}{2}\, \underline{y}^T(t)\underline{P}\,\underline{y}(t), \tag{6.66}$$

and therefore the following integral equation is obtained for the adjoint row vector $\underline{\psi}^T(\tau)$:

$$\underline{\psi}^T(\tau) + \underline{y}^T(\tau)\underline{P} = \int_{t=\tau}^{T} \underline{\psi}^T(t)\underline{\tilde{G}}(t-\tau)[\tfrac{1}{2}(\underline{K}_1+\underline{K}_2) +$$

$$+ \frac{1}{2}(\underline{K}_1-\underline{K}_2)\tilde{u}(\tau)]dt. \tag{6.67}$$

The $\tilde{u}$-dependent part of the Π-function now takes the form

$$\int_{\tau=t}^{T} \underline{\psi}^T(\tau)\underline{\tilde{G}}(\tau-t)\tfrac{1}{2}(\underline{K}_1-\underline{K}_2)\underline{y}(t)\tilde{u}d\tau \overset{!}{=} \max_{\tilde{u}\in[-1,1]} \tag{6.68}$$

and therefore the optimal $\tilde{u}(t)$ must satisfy

$$\tilde{u}(t) = \mathrm{sgn}[\int_{\tau=t}^{T} \underline{\psi}^T(\tau)\underline{\tilde{G}}(\tau-t)d\tau\cdot(\underline{K}_1-\underline{K}_2)\underline{y}(t)]. \tag{6.69}$$

Eqs.(6.59), (6.67) and (6.69) have to be solved simultaneously to obtain the optimal switching function $\tilde{u}(t)$, which may, of course, pose non-trivial computational problems.

In order to get some more insight into the nature of optimal variable structure control, the special case of scalar output $y(t)$ will be considered in the following. The matrix $\underline{P}$ in Eq. (6.62) now degenerates to a number $p > 0$, and can therefore be specified arbitrarily, e.g. $p = 1$. Hence, Eqs. (6.59), (6.67) and (6.69) will take the form

$$y(t) = \frac{1}{2}\int_{\tau=0}^{t} \tilde{g}(t-\tau)[(k_1+k_2)+(k_1-k_2)\tilde{u}(\tau)]y(\tau)d\tau + y_o(t), \tag{6.70}$$

$$\psi(\tau) + y(\tau) = \frac{1}{2}[(k_1+k_2)+(k_1-k_2)\tilde{u}(\tau)]\cdot$$

$$\cdot\int_{t=\tau}^{T} \tilde{g}(t-\tau)\psi(t)dt, \tag{6.71}$$

$$\tilde{u}(t) = \mathrm{sgn}[(k_1-k_2)y(t)\int_{\tau=t}^{T} \tilde{g}(\tau-t)\psi(\tau)d\tau]. \tag{6.72}$$

From Eq.(6.72) it can be observed that the switchings of $\tilde{u}(t)$ are the zeros of the output $y(t)$ and those of the integral

$$\int_{\tau=t}^{T} \tilde{g}(\tau-t)\Psi(\tau)d\tau .$$

6.3.3 Computational aspects and numerical example

Similar to the considerations of section 6.2.5, the adjoint integral equation (6.71) is rewritten in the form

$$\Psi_{k+1}(t) = - y_k(t) + \frac{1}{2}[(k_1+k_2) + (k_1-k_2)\tilde{u}(t)]\cdot$$

$$\cdot\int_{\tau=t}^{T} \tilde{g}(\tau-t)\Psi_k(\tau)d\tau \qquad (6.73)$$

for the purpose of iterative computation, index k denoting again the number of iteration steps.

The basic signal flow diagram thus obtained for the iterative solution of Eqs.(6.70), (6.71) and (6.72) is very similar to that of Fig.6.2 and therefore need not be repeated here. Again the hybrid computer device proposed in [42] has been employed to compute the following numerical example.

Example 6.2:

Let the linear distributed plant to be controlled be given by the heat equation

$$\frac{\partial x(t,z)}{\partial t} = \frac{\partial^2 x(t,z)}{\partial z^2}, \qquad 0 \leq z \leq 1, \qquad (6.74)$$

with boundary conditions

$$\left.\frac{\partial x}{\partial z}\right|_{z=0} = - u_1(t), \qquad \left.\frac{\partial x}{\partial z}\right|_{z=1} = 0, \qquad (6.75)$$

where $u_1(t)$ is a scalar control variable, and initial condition

$$x(0,z) = x_o(z), \qquad (6.76)$$

and let the scalar output be defined by

$$y(t) = x(t,0). \tag{6.77}$$

The lumped parameter subsystem in Fig.4.4 may be specified by the scalar state equation

$$\dot{\xi}(t) = -5\xi(t) + 5[-k(t)y(t)], \tag{6.78}$$

with initial condition

$$\xi(0) = \xi_0 \tag{6.79}$$

and with relation

$$u_1(t) = \xi(t). \tag{6.80}$$

The gain $k_1 = 40$ has been specified by frequency domain methods such as to make the linear control-loop asymptotically stable for this k_1. The alternative gain k_2 has been set equal to zero.

In the procedure for the computation of the optimal switching function $\tilde{u}(t)$ the distributed plant has been approximated by series truncation, as the system exhibits spatial low-pass nature to a high degree.

Fig.6.7 outlines the numerical result obtained for initial perturbation

$$x_0(z) \equiv 1, \quad \xi_0 = 0,$$

and terminal time T = 0,5. As can be seen from curve (a), the damping ratio of the linear over-all system with gain $k_1 = 40$ is rather small. In the linear case with $k = (k_1+k_2)/2 = 20$ the overshoot has of course decreased but on the other hand the peak time has increased (curve(b)). As curve (c) indicates, optimal variable structure control offers better performance than any linear regulator. This is essentially the same result as obtained by Becker [2] in the lumped parameter case.

As a critical comment it should be stressed that the optimal switching function $\tilde{u}(t)$ in the above procedure is not obtained

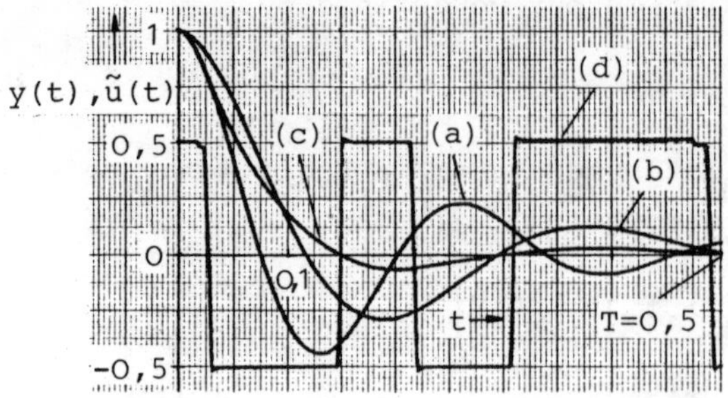

Fig.6.7: Comparison of performance with fixed linear control and optimal variable structure control.
(a) output y(t) in case of linear control with fixed $k(t) \equiv k_1 = 40$
(b) output y(t) in case of linear control with fixed $k(t) \equiv (k_1+k_2)/2 = 20$
(c) output y(t) in case of optimal variable structure control with

$$k(t) = \begin{cases} k_1 = 40 \text{ if } \tilde{u}(t) = +1 \\ k_2 = 0 \text{ if } \tilde{u}(t) = -1 \end{cases}$$

(d) optimal switching function $\tilde{u}(t)/2$

in terms of the actual state of the system but in terms of the initial state. Therefore, $\tilde{u}(t)$ must be computed in advance, before starting the control process. More effort seems to be worth-while to derive near-optimal switching strategies in terms of some on-line measured state or output variables. In total, design procedures basing on Ljapunow's direct method may prove to be more feasible than those basing on maximum principles.

7. Concluding remarks

This contribution has presented a first approach to bilinear distributed parameter systems and does of course not claim to be an exhaustive study. The goal was to include and illustrate the wide range of problems which can be modelled by partial differential equations with parametric control variables. It has been tried to motivate both applicative and theoretical interest in this new class of dynamical systems.

From the applicative point of view, special emphasis has been given to problems of transportation and to questions of heat and mass transfer. Systems with a moving boundary, which can be modelled as a special class of bilinear distributed systems, have just been touched and offer a wide field for future research, especially in biological population dynamics.

From the theoretical point of view, stability, near-optimal and optimal control of distributed systems including multiplicative control variables, have been studied to some extent. Ljapunow's direct method and Butkovskiy's maximum principle have been employed as mathematical tools for the design of stable closed-loop systems and optimal control systems, respectively.

More research work seems to be worth-wile in the field of structural properties. There is a lack of necessary and sufficient conditions of reachability and controllability especially in case of bounded control. Also the benefit of independent linear and multiplicative control variables needs further investigation. Questions of state observability and related problems of optimal sensor location, which have been given fairly exhaustive answers for linear distributed systems in recent years, are an appealing challenge in case of bilinear distributed systems.

Some of the theoretical methods for the control of distributed parameter systems do not provide explicit feedback control laws but a set of very complex partial differential or integral equations. Therefore, real time implementation needs larger research effort in the development of computational algorithms, particularly with regard to the now available microcomputers.

It is the author's conviction that bilinear distributed parameter models can play a key role also in the design of highly complex variable structure control systems which have not been treated in this contribution. For example, bilinear models may prove to be an effective tool for the treatment of systems with redundancy, hierarchical systems and large scale systems.

References

Bilinear lumped parameter systems:

[1] Becker, C.: Beschreibung strukturvariabler Regelungssysteme als eine spezielle Klasse bilinearer Systeme. Regelungstechnik 25 (1977), pp. 364 - 366.

[2] Becker, C.: Synthese strukturvariabler Regelungssysteme mit Hilfe von Ljapunow-Funktionen. Stuttgart: Hochschulverlag 1979.

[3] Buyakas, V.I.: Optimal Control by Systems with Variable Structure. Automat. Telemekh. vol. 27, no.4 (1966).

[4] Bruni, C.; Di Pillo, G.; Koch, G.: Bilinear Systems, an Appealing Class of "Nearly Linear" Systems in Theory and Applications. IEEE Trans. on Automatic Control AC - 19 (1974), pp. 334 - 348.

[5] Emeljanov, S.V.: Automatische Regelungssysteme mit veränderlicher Struktur. München and Wien: R. Oldenbourg Verlag 1969.

[6] Hofer, E.; Sagirow, P.: Optimal Systems Depending on Parameters. AIAA Journal, Vol.6 (1968), pp. 953 - 956.

[7] Hofer, E.P.: Zur Theorie und Anwendung optimaler bilinearer Systeme, Dissertation, TU Stuttgart 1970.

[8] Itschner, B.: Eine suboptimale Schaltstrategie für höherwertige rechenzeitsparende DDC-Algorithmen. Regelungstechnik 25 (1977), pp. 142 - 144.

[9] Kiendl, H.: Suboptimale Regler mit abschnittsweise linearer Struktur. Berlin: Springer-Verlag 1972.

[10] Mohler, R.R.; Rink, R.E.: Multivariable Bilinear System Control. Proceedings of the IFAC-Symposium on Multivariable Control Systems, Düsseldorf 1968.

[11] Mohler, R.R.; Rink, R.E.: Control with a Multiplicative Mode. J. Basic Eng., Trans ASME 91 (1969), pp.201 - 206.

[12] Mohler, R.R.: Bilinear Control Processes. New York: Academic Press 1973.

[13] Mohler, R.R.; Ruberti, A. (Editors): Variable Structure Systems with Application to Economics and Biology. Proceedings of the second US-Italy Seminar on Variable Structure Systems 1974. New York: Springer-Verlag 1975.

[14] Ruberti, A.; Isidori, A.; D'Alessandro, P.: Theory of Bilinear Dynamical Systems. New York: Springer-Verlag 1972.

[15] Williamson, D.: Observation of Bilinear Systems with Application to Biological Control. Automatica, Vol. 13 (1977), pp. 243 - 254.

<u>Introductory and advanced textbooks and surveys on distributed parameter systems</u>:

[16] Brogan, W.L.: Optimal Control Theory Applied to Systems Described by Partial Differential Equations; in: Advances in Control Systems 6. New York and London: Academic Press 1968.

[17] Butkovskiy, A.G.: Distributed Control Systems. New York: American Elsevier Publ.Comp. 1969.

[18] Gilles, E.D.: Systeme mit verteilten Parametern. München and Wien: R. Oldenbourg Verlag 1973.

[19] Goodson, R.E.; Polis, M.P.: A Survey of Parameter Identification in Distributed Systems. 6. IFAC-Congress, Boston/USA 1975, paper 8.2.

[20] Lions, J.L.: Optimal Control of Systems Governed by Partial Differential Equations. New York: Springer-Verlag 1971 (French original: Dunod 1968).

[21] Ray, W.H.; Lainiotis, D.G. (Editors): Distributed Parameter Systems; Identification, Estimation, and Control. New York and Basel: Marcel Dekker, Inc. 1978.

[22] Robinson, A.C.: A Survey of Optimal Control of Distributed Parameter Systems. Automatica 7 (1971), pp.371-388.

[23] Ruberti, A. (Editor): Distributed Parameter Systems: Modelling and Identification. Proceedings of the IFIP Working Conference, Rome/Italy 1976. New York: Springer-Verlag 1978.

[24] Wang, P.K.C.: Control of Distributed Parameter Systems; in: Advances in Control Systems 1. New York and London: Academic Press 1964.

<u>Mathematical textbooks on partial differential equations and functional analysis:</u>

[25] Courant, R.; Hilbert, D.: Methoden der mathematischen Physik, Vol. I and II. New York: Springer-Verlag, 3rd edition 1968.

[26] Frank, Ph.; Mises, R.:Differentialgleichungen der Physik, Vol. I and II. Braunschweig: Vieweg Verlag 1961.

[27] Großmann, S.: Funktionalanalysis, Vol. I and II. Leipzig: Akademische Verlagsgesellschaft 1970.

[28] Kreyszig, E.: Introductory Functional Analysis with Applications. New York: J. Wiley 1978.

[29] Morse, Ph. M.; Feshbach, H.: Methods of Theoretical Physics, Vol. I and II. New York: McGraw-Hill 1953.

[30] Smirnow, W.L.: Lehrgang der höheren Mathematik, Vol. IV and V. Berlin: VEB Deutscher Verlag der Wissenschaften, 4th edition 1966.

[31] Sommerfeld, A.: Partielle Differentialgleichungen der Physik. Leipzig: Akademische Verlagsgesellschaft, 6th edition 1966.

[32] Wulich, B.S.: Einführung in die Funktionalanalysis, Vol. I and II. Stuttgart: Teubner Verlagsgesellschaft 1961/62.

Bilinear distributed parameter systems:

[33] Balakrishnan, A.V.: Stochastic Bilinear Partial Differential Equations. In [13], pp. 1 - 43.

[34] Briggs, D.L.; Shen, C.N.: Switching Analysis for Constrained Bilinear Distributed Parameter Systems with Applications. J. Basic Eng., Trans ASME 91 (1969), pp. 277 - 283.

[35] Bruni, C.; Koch, G.: A Degenerate (Bounded Rate) Class of Distributed Parameter Systems. In [23], pp. 138 - 152.

[36] Franke, D.: Ljapunow-Synthese linearer konzentrierter Regler für bilineare örtlich verteilte Strecken. Regelungstechnik 27 (1979), pp. 213 - 220.

[37] Franke, D.: Optimal Bilinear Boundary Control of a Distributed Parameter Cooling Process. Preprints of the 2nd IFAC-Symposium on Optimization Methods, Varna/Bulgaria 1979.

[38] Koch, G.: A Realization Theorem for Infinite Dimensional Bilinear Systems. Ricerche di Automatica, Vol.3 (1972).

Various references:

[39] Butkovskiy, A.G.: The Method of Moments in the Theory of Optimal Control of Systems with Distributed Parameters. Aut. Rem. Control 24, No.9 (1963), pp. 1106 - 1113.

[40] Corduneanu, C.: Integral Equations and Stability of Feedback Systems. New York: Academic Press 1973.

[41] Föllinger, O.: Nichtlineare Regelungen III: Ljapunow-Theorie und Popow-Kriterium. München and Wien: R. Oldenbourg Verlag 1970.

[42] Franke, D.: Optimierung dynamischer Systeme durch Lösen der Butkovskiyschen Integralgleichungen mittels iterativer Rechenschaltungen. VDI-Forschungsheft 568. Düsseldorf: VDI-Verlag 1975.

[43] Hoppenstaedt, F.C.: Optimal Exploitation of a Spatially Distributed Fishery. In "New Trends in Systems Analysis", Edited by A. Bensoussan and J.L. Lions. New York: Springer-Verlag 1977.

[44] Kalman, R.E.: On the General Theory of Control Systems. Proceedings of the First International Congress on Automatic Control, Moskau 1960. Butterworths 1961, vol.1, pp. 481 - 492.

[45] Kalman, R.E.; Bertram, J.E.: Control System Analysis and Design Via the "Second Method" of Ljapunow. J. Basic Eng., Trans ASME 82 (1960), pp. 371 - 400.

[46] Köhne, M.: Zustandsbeobachter für Systeme mit verteilten Parametern - Theorie und Anwendung. Fortschrittberichte der VDI-Zeitschriften, Reihe 8, Nr. 26. Düsseldorf: VDI-Verlag 1977.

[47] Köhne, M.: The Control of Vibrating Elastic Systems. In [21], pp. 387 - 456.

[48] La Salle, J.; Lefschetz, S.: Die Stabilitätstheorie von Ljapunow, die direkte Methode mit Anwendungen. Mannheim: Bibliographisches Institut, Hochschultaschenbücher-Verlag 1967.

[49] Massera, J.L.: Contributions to Stability Theory. Ann. Math., Vol. 64, 1956, pp. 182 - 206.

[50] Ockendon, J.R.; Hodkins, W.R. (Editors): Moving Boundary Problems in Heat Flow and Diffusion. Oxford: Clarendon Press 1975.

[51] Parks, P.C.; Pritchard, A.J.: On the Construction and Use of Ljapunow Functionals. 4th IFAC-Congress, Warschau 1969, paper No. 20.5.

[52] Persidskii, K.: On the Stability of Solutions of Denumerable Systems of Differential Equations. Akademii Nauk Kazakhskio SSSR, Ser. Matem., Vol.2, No.1, 1948, pp.2-35.

[53] Schober, J.: Anwendung funktionalanalytischer Methoden zur Stabilitätsanalyse von Systemen mit verteilten Parametern. Dissertation, Universität Karlsruhe 1978.

[54] Schrodi, E.: Optimale Steuerung linearer dynamischer Systeme durch Lösen Fredholmscher Integralgleichungen. Stuttgart: Hochschulverlag 1979.

[55] Venkatesh. Y.V.: Energy Methods in Time-Varying System Stability and Instability Analyses. New York: Springer-Verlag 1977.

[56] Wang, P.K.C.: Stability Analysis of Elastic and Aeroelastic Systems via Ljapunow's Direct Method. Journal of The Franklin Institute, Vol. 281, No.1,(1966), pp.51-72.

[57] Wang, P.K.C.: On the Feedback Control of Distributed Parameter Systems. Int.J.Control, 1966, Vol.3, No.3, pp. 255 - 273.

[58] Wang, P.K.C.: Asymptotic Stability of Distributed Parameter Systems with Feedback Controls. IEEE Trans. Aut. Control, Vol. AC-11, No.1 (1966), pp. 46 - 54.

[59] Wang, P.K.C.: On a Class of Optimization Problems Involving Domain Variations. In "New Trends in Systems Analysis", Edited by A.Bensoussan and J.L. Lions. New York: Springer-Verlag 1977.

[60] Zeitz, M.: Nichtlineare Beobachter für chemische Reaktoren. Fortschrittberichte der VDI-Zeitschriften, Reihe 8, Nr. 27. Düsseldorf: VDI-Verlag 1977.

[61] Zubov, V.I.: Methods of A.M. Ljapunov and their Applications. Publ. House of Leningrad Univ. 1957.

Index

Contribution II

Part I: A Generalisation of the Kalman-Filter-Algorithm and its Application in Optimal Identification

Part II: State and Parameter Estimation of Linear Systems in Arbitrary State Coordinates

Erhard Bühler

Contents

Contribution II

Erhard Bühler

Notations

$\underline{A}$ $\underline{B}$ $\underline{C}$ $\underline{H}$ $\underline{\Phi}$ $\underline{\Gamma}$	- system matrices
c c'	- scaling factor
det	- determinant
$\underline{E}$ $\underline{e}$	- identity matrix (vector)
g(·)	- weighting function
m	- dimension of $\underline{\gamma}$
ML MAP	- Maximum-likelihood Maximum-a-posteriori
$\underline{Q}$ q $\underline{R}$ r	- noise covariances
q	- dimension of $\underline{\delta}$
p(·) p(·|·)	- density conditional density
$\underline{P}$	- covariance matrix
RLQ	- Recursive Least Square Method
t(·)	- estimation function
$\underline{T}$	- transformation matrix
$\underline{u}$ $\underline{y}$	- measurable input measurable output
$\underline{w}$ $\underline{v}$	- unmeasurable noise (vectors)
$\underline{x}$ $\underline{z}$	- states (vectors)
$\underline{\gamma}$ $\underline{\delta}$ $\underline{a}$ $\underline{b}$	- unknown parameters
$\underline{\psi}$	- map, defining the relation between $\underline{\delta}$ and $\underline{\gamma}$
Ω Σ $\tilde{\Sigma}$	- parameter sets
$\|\ \|$	- norm
$\wedge$	- e.g. $\hat{x}$ = estimate of x
$\sim$	- e.g. $\tilde{x}$ = estimation error of x
$*$	- e.g. x^* = augmented state vector
$-$	- e.g. $\bar{\delta}$ = estimate, based on simplified assumptions

PART I: A Generalisation of the Kalman-Filter-Algorithm and its Application in Optimal Identification

Summary

Despite the advances in the areas of processing and memory techniques, primarily simple, numerically robust and easily definable methods have been favoured in on-line identification. The principle presented in this paper for combined estimation of state and parameters of linear dynamic systems also has these characteristics. On the one hand, this method is applicable as an adaptive state estimator, on the other hand, because of its small computational effort, the principle used as a parameter estimator presents an alternative to the well known "Recursive Least Squares Method".

I.1 Introduction

Many methods of system identification have been developed and published in past years. Most of them are based on linear or linearizable models, since otherwise the methods of solution would be too extensive and consequently would only find practical application in special cases.

Whereas with the problem of parameter identification further new algorithms can be expected, the theory of linear state estimation, which has increasing importance in modern control theory, is essentially summarized by the work of Kalman and Bucy /10/, /11/. The application of the Kalman-Bucy-Filter requires that all system parameters have to be known, which is a disadvantage that justifies the development of adaptive state-estimators.

The new principle presented in section 3 and 4 and the resulting algorithm serve for combined or simultaneous estimation of states of linear dynamic systems.

For this recursive and therefore on-line-suitable method exist two application areas. On one hand, it may be defined as an adaptive state estimator for a large class of linear systems having unknown parameters. In this case, simplicity of solution is especially emphasized, since the state-parameter estimation is non-linear and therefore represents a very complicated estimation problem /5/. On the other hand, the application is suitable for those problems where only parameter estimates are of interest. For such cases, an extra calculation of the state estimates can be of advantage in the parameter estimation, although the state estimates are not of primary interest. This becomes especially apparent by comparison with one of the most popular on-line parameter estimation techniques, the so called "Recursive Least Squares Method" (RLQ) /1/,/6/, whose estimates as a simple special case can be calculated using the new state-parameter estimation principle, but which can be applied to more general models.

Worth particular attention is the definite advantage of the relatively small computational requirements, which in the worst case, including state estimation and calculation of variances, is approximately 1.5 times greater than required for the RLQ (simplest version).

The state-parameter estimation principle is based on recognition of the fact that for a certain class of linear systems, the optimal solution for the inevitable non-linear estimation problem is simply definable. The optimization for which verification is given by generalization of the Kalman-Bucy-Theorem is restricted to systems containing input noise but no measurement noise. In cases where additional measurement noise is present, a suboptimal method follows from a minor modification of the original results. It is shown that this method provides excellent results, particularly in parameter estimation problems.

I.2 Statement of the problem

Since digital computers are used in most cases for on-line identification, only time-discrete systems will be considered in the following, although on principle the transfer of the results to continuous systems is possible.

As previously mentioned, the problem of combined estimation of the states and parameters of linear systems is generally not exactly solvable without resorting to extensive, and therefore for on-line identification, unacceptable methods. However this is not valid for a certain class of systems for which a very simple optimal solution exists.

Starting point is the following system type of order n where an assumed measureable input u_k and output y_k are introduced (for vectored input or output see /4/).

$$\underline{x}_{k+1} = (\underline{A}_k + \tilde{\underline{A}}_k)\underline{x}_k + \underline{b}_k u_k + \underline{w}_k \qquad (1)$$

$$y_k = \tilde{\underline{c}}_k^T \underline{x}_k \qquad (2)$$

where $\tilde{\underline{A}}_k = \left[\underline{0} \cdots \underline{a}_k \cdots \underline{0} \right]$, $\underline{a}_k = \underline{M}_k^a \, \underline{\delta}_k$

$\tilde{\underline{c}}_k^T = \left[0 \cdots c_k \cdots 0 \right]$, $\underline{b}_k = \underline{M}_k^b \, \underline{\delta}_k + \underline{m}_k^b$

and $$\underline{\delta}_{k+1} = \underline{\Theta}_k \, \underline{\delta}_k + \underline{w}_k^{\delta} \qquad (3)$$

$\underline{A}_k$, $\underline{M}_k^a$, c_k, $\underline{M}_k^b$, $\underline{m}_k^b$, $\underline{\Theta}_k$ are arbitrary, known values. Unknowns are the state vector $\underline{x}_k$ and the parameter vector $\underline{\delta}_k$ which, for the sake of the following analysis, can be combined to an augmented vector:

$$\underline{x}_k^* := \begin{bmatrix} \underline{x}_k \\ \\ \underline{\delta}_k \end{bmatrix}$$

$\underline{w}_k^x$ and $\underline{w}_k^\delta$ are white Gaussian noise processes with

$$\underline{w}_k^* = \begin{bmatrix} \underline{w}_k^x \\ \\ \underline{w}_k \end{bmatrix} \sim N(\underline{0},\underline{Q}_k^*).$$

Further given is $\underline{x}_o^* \sim N(\overline{\underline{x}}_o^*,\overline{\underline{P}}_o^*)$. In the matrix $\underline{\tilde{A}}$ there is exactly one column not equal to zero. At the same position (column index) the row vector $\underline{\tilde{c}}_k^T$ contains an element different from zero. In equation (3) (parameter model) the case of constant parameters is included via $\underline{\Theta}_k = \underline{E}$ and $\underline{w}_k^\delta \equiv \underline{0}$.

An essential restriction, which will be discarded in section 4 is that in the above system only input disturbances and no measurement disturbances are permitted. The special manner of defining the unknown parameters in the system equations is less restrictive, because all observable systems /7/ - and only such are of interest for the current problem statement - can be transformed to the above type of system. This is obvious since equations (1) and (2) contain the important special case of the canonical form

$$\underline{x}_{k+1} = \underline{A}\ \underline{x}_k + \underline{b}\ u_k \tag{4}$$

$$y_k = \begin{bmatrix} 0 \ldots 0 1 \end{bmatrix} \underline{x}_k \tag{5}$$

where $\underline{A} = (\underline{A}_k + \underline{\tilde{A}}_k) = \left[\begin{array}{c|c} 0\ldots 0 & \\ \hline \underline{E} & \underline{a} \end{array}\right]$, $\underline{\delta}_k = \underline{\delta} = \begin{bmatrix} \underline{a} \\ \hline \underline{b} \end{bmatrix}$

In the following, this form will be defined as the observer form. Note that there is a further canonical form which contains the matrix $\underline{A}$ in the transposed form and which, in relation to the state-parameter-estimation problem, has a fully different character and is not suitable in the way of solution being considered here.

Example:

Given the spring-mass system (e.g. conveyer-basket on elastic cable) shown in figure 1

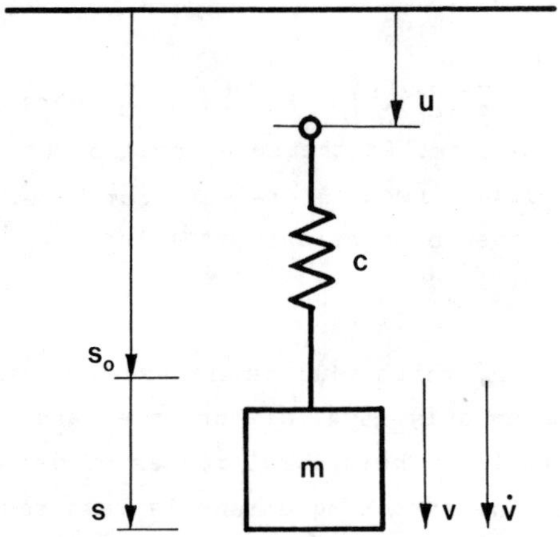

Figure 1: Spring-Mass System

$$\dot{s} = v$$
$$\dot{v} = -\frac{c}{m}\ s + \frac{c}{m}\ u$$

Measured values: $y_k = s(t_k)$

Unknown values: $v(t_k)$ and $\omega = \frac{c}{m}$

Discretization of the system and then transforming to the observer form gives

$$\underline{x}_{k+1} = \begin{bmatrix} 0 & -1 \\ 1 & 2\cos\omega T \end{bmatrix} \underline{x}_k + \begin{bmatrix} \cos\omega T - 1 \\ 1 - \cos\omega T \end{bmatrix} u_k$$

$$y_k = \begin{bmatrix} 0 & 1 \end{bmatrix} \underline{x}_k \ .$$

It is reasonable here not to define the elements $\underline{a}$ and $\underline{b}$ as four unknown free parameters but to use the simple common link between these elements by defining $\delta = \cos\omega T$. In this way, a more effective estimation results from using only one unknown parameter. With $\delta = \cos\omega T$ the system belongs to the class of systems defined by (1) and (2) above, whereby

$$\underline{A}_k = \begin{bmatrix} 0 & -1 \\ 1 & 0 \end{bmatrix}, \qquad \underline{a}_k = \begin{bmatrix} 0 \\ 2 \end{bmatrix} \delta$$

$$\underline{\tilde{c}}_k^T = \begin{bmatrix} 0 & 1 \end{bmatrix}, \qquad \underline{b}_k = \begin{bmatrix} 1 \\ -1 \end{bmatrix} \delta + \begin{bmatrix} -1 \\ 1 \end{bmatrix} .$$

Generally, the problem is the calculation of the distribution $p(\underline{x}_k^* | \underline{y}^k)$ for the state-parameter-vector $\underline{x}_k^*$ knowing the measured values $\underline{y}^k = \begin{bmatrix} y_0 & \dots & y_k \end{bmatrix}^T$ or the computing of a suitable estimate $\hat{\underline{x}}_k^*$. The following technique provides, in a simple way, both.

I.3 The optimal solution

The augmented vector $\underline{x}_k^* = \left[\underline{x}_k^T, \underline{\delta}_k^T\right]^T$ can be interpreted as the state-vector of a non-linear system. Although numerous representations are possible for such a system, only the following is of significance. Through suitable manipulation of (1), (2) and (3) above it can be established:

$$\underline{x}_{k+1}^* = \left[\begin{array}{c|c} \underline{A}_k & \frac{y_k}{c_k}\,\underline{M}_k^a + u_k\,\underline{M}_k^b \\ \hline \underline{0} & \underline{\Theta}_k \end{array}\right] \underline{x}_k^* + \left[\begin{array}{c} \underline{m}_k^b \\ \\ \underline{0} \end{array}\right] u_k + \left[\begin{array}{c} \underline{w}_k^x \\ \\ \underline{w}_k^\delta \end{array}\right] \tag{6}$$

$$y_k = \left[\begin{array}{c|ccc} \underline{c}_k^T & 0 & \cdots & 0 \end{array}\right] \underline{x}_k^* \; . \tag{7}$$

The special case, the observer form having constant parameters $\underline{\delta}^T = \left[\, \underline{a}^T, \underline{b}^T \right]$, results in

$$\underline{x}_{k+1}^* = \left[\begin{array}{cc|c|c} \begin{array}{c} 0 \cdots \\ \underline{E} \end{array} & \begin{array}{c} 0 \\ \vdots \\ 0 \end{array} & y_k\,\underline{E} & u_k\underline{E} \\ \hline \underline{0} & & \underline{E} & \end{array}\right] \underline{x}_k^* + \left[\begin{array}{c} \underline{w}_k^x \\ \\ \underline{0} \end{array}\right] \tag{8}$$

$$y_k = \left[0 \cdots 0\,1\;0 \cdots 0\right] \underline{x}_k^* \; . \tag{9}$$

In shortened form, the extended system equations are

$$\underline{x}_{k+1}^* = \underline{A}_k^* \; \underline{x}_k^* + \underline{m}_k^* \, u_k + \underline{w}_k^* \tag{10}$$

$$y_k = \underline{c}_k^{*T} \; \underline{x}_k^* \; . \tag{11}$$

As we can see, for such system types the Kalman-Bucy-Filter equations for estimating $\underline{x}_k^*$ can be formally applied.

Nevertheless, the following has to be noticed:

- $\underline{A}_k^*$ is a random matrix since it is dependent on y_k
- the system is non-linear
- a non-linear estimation results
- the output signal y_k is not Gaussian.

This means that the assumptions, well-known from the literature (e.g. /10/) for the Kalman-Bucy-Filter equations, are not fulfilled. Clarity is given by the following Theorem which presents a generalisation for the Kalman-Bucy-Filter Theorem.

Theorem 1:

Given the non-linear stochastic system of order p

$$\underline{z}_{k+1} = \underline{\Phi}(k,\underline{y}^k)\, \underline{z}_k + \underline{f}(k,\underline{y}^k) + \underline{\Gamma}^1(k,\underline{y}^k)\, \underline{w}_k \qquad (12)$$

$$\underline{y}_k = \underline{C}(k,\underline{y}^{k-1})\underline{z}_k + \underline{g}(k,\underline{y}^{k-1}) + \underline{\Gamma}^2(k,\underline{y}^{k-1})\underline{v}_k \qquad (13)$$

where $\underline{y}^k = \left[\, y_o \cdots y_k \right]^T$

or abbreviated:

$$\underline{z}_{k+1} = \underline{\Phi}_k\, \underline{z}_k + \underline{f}_k + \underline{\Gamma}_k^1\, \underline{w}_k \qquad (14)$$

$$\underline{y}_k = \underline{C}_k\, \underline{z}_k + \underline{g}_k + \underline{\Gamma}_k^2\, \underline{v}_k \qquad (15)$$

$\underline{w}_k$ and $\underline{v}_k$ are white independent Gaussian processes with mean $\underline{0}$ and covariance $\underline{Q}_k$ and $\underline{R}_k$. $\underline{z}_o$ is also Gaussian with $\underline{z}_o \sim N(\overline{\underline{z}}_o, \overline{\underline{P}}_o)$. Then, the a posteriori distribution $p(\underline{z}_k | \underline{y}^k)$ is Gaussian with the monents

$$\hat{\underline{z}}_{k|k} = E(\underline{z}_k \mid \underline{y}^k) \quad ,$$

$$\underline{P}_{k|k} = E(\tilde{\underline{z}}_k \tilde{\underline{z}}_k^T \mid \underline{y}^k) = \operatorname{cov}(\underline{z}_k, \underline{z}_k \mid \underline{y}^k) \text{ with } \tilde{\underline{z}}_k = \underline{z}_k - \hat{\underline{z}}_{k|k} \quad ,$$

that can be calculated via the Kalman-Bucy-Filter Equations:

$$\hat{\underline{z}}_{k+1|k+1} = \hat{\underline{z}}_{k+1|k} + \underline{K}_{k+1}(\underline{y}_{k+1} - \hat{\underline{y}}_{k+1|k}) \tag{16}$$

$$\underline{P}_{k+1|k+1} = \langle \underline{E} - \underline{K}_{k+1}\underline{C}_{k+1} \rangle \underline{P}_{k+1|k} \tag{17}$$

where

$$\underline{K}_{k+1} = \underline{P}_{k+1|k}\, \underline{C}_{k+1}^T \Big\langle \underline{\Gamma}_{k+1}^2 \underline{R}_{k+1} \underline{\Gamma}_{k+1}^{2T} + \underline{C}_{k+1}\, \underline{P}_{k+1|k}\, \underline{C}_{k+1}^T \Big\rangle^{-1} \tag{18}$$

$$\hat{\underline{y}}_{k+1|k} = \underline{C}_{k+1}\, \hat{\underline{z}}_{k+1|k} + \underline{g}_{k+1} \tag{19}$$

$$\hat{\underline{z}}_{k+1|k} = \underline{\Phi}_k \hat{\underline{z}}_{k|k} + \underline{f}_k \tag{20}$$

$$\underline{P}_{k+1|k} = \underline{\Phi}_k \underline{P}_{k|k} \underline{\Phi}_k^T + \underline{\Gamma}_k^1 \underline{Q}_k \underline{\Gamma}_k^{1T} \tag{21}$$

and the Initial values

$$\hat{\underline{z}}_{o|-1} = \overline{\underline{z}}_o \,,\; \underline{P}_{o|-1} = \overline{\underline{P}}_o,\; \hat{y}_{o|-1} = \underline{C}_o \overline{\underline{z}}_o + \underline{g}_o \tag{22}$$

<u>Proof: See appendix</u>

The following should be noted in connection with this result. The equations (12) and (13) present a generalized system in comparison with equations (6) and (7). In this way, for example, all system matrices and vectors can be dependent on the observations. The noise $\underline{v}_k$ is only introduced to generalize the problem and is not considered in the suggested principle. The resulting singularity is however, for discrete systems unproblematical - in (18) $\underline{R}_{k+1} = \underline{0}$ merely needs to be introduced /2/.

Note that the conditional mean as estimate always leads to the least-square error.

Theorem 1 shows that the optimal solution for the non-linear estimation problem is provided by the formal application of the Kalman-Bucy-Filter-algorithm. Although under conventional assumptions the Kalman-Bucy-Filter leads to a linear estimate /13/, in the following, the algorithm defined by Theorem 1 will be called the Non-Linear Kalman-Bucy-Filter (NKBF).

The formal relation to the (linear) Kalman-Bucy-Filter should not distract us from the fact that the basic linear characteristic of the estimate cannot be transferred to the non-linear problem. E.g. the simple equations of sensitivity analysis known for the linear case /9/ do not hold with respect to NKBF. On the other hand, in the case where the disturbances are not Gaussian the NKBF can be applied with the same justification, as is often done in the linear case (state estimation with known parameters).

The solution of the optimal smoothing problem is given by formally transferring the results of the linear case to the non-linear case /4/. The corresponding optimal prediction is only possible for one step, however.

The optimal principle is also applicable to continuous systems, but the singularity, owing to the zero measurement noise, is more difficult than for discrete systems.

Consideration of more than one input is not problematical and leads only to a formal extension. For systems with more than one output, attention has to be given to the fact that first, the observer form is not unique, and second, that the output matrix may contain unknown parameters so that Theorem 1 can be applied only in a restricted way.

Although the theory contains several significant theorectical aspects, in the following, the emphasis is placed on its technique of application.

The effort of calculation associated with the NKBF is relatively small. This is especially the case with the observer form since in such cases few matrix elements differ from 0 or 1. By appropriate programming and use of symmetry characteristics, one requires approximately $10(n^2+n)$ multiplications for a system of order n (single input, single output). The number of additions is of the same order. This is only about 1.5 times more than is necessary with the "Recursive Method of Least Squares" (RLQ). This method however provides only parameter estimates. For systems with more inputs the relationship is even better.

As mentioned initially, this relatively small expense associated with the NKBF-algorithm will also lead to application in cases where only parameter estimates are of interest, since the introduction of state values can bring certain advantages in realising a parameter estimation. This is especially seen in comparison with RLQ whose estimates, as a special case, result from the NKBF although both algorithms essentially differ. This situation arises from the fact that the RLQ estimate results in the maximum-likelihood-estimate $\hat{\underline{\delta}}_k^{ML}$ for certain models (independent Gaussian residuals). This estimate is identical with the conditional mean $\hat{\underline{\delta}}_{k \mid k} = E(\underline{\delta} \mid \underline{y}^k)$ determined from the NKBF when applying a Gaussian a priori distribution with a sufficiently large variance. In practice, this can be realized unproblematicly.

The important characteristics of the NKBF, especially in comparison with the RLQ technique, are presented in short summary form in the following:

- The NKBF presents an uncomplicated, easily definable on-line technique, which is also applicable to models for which the RLQ and other methods are either not applicable or require extensive and complicated modifications.

- Without modification the NKBF is valid for continuous systems.

- The NKBF allows inclusion of a priori knowledge about $\underline{x}_o$ as well as the correlation of $\underline{x}_o$ and $\underline{\delta}_o$.

- Input disturbances can be allowed in a more versatile manner with the NKBF technique than is possible with RLQ. In special cases, prewhitening of the inputs and output can be helpful by RLQ but this approach is only practicable in cases with constant parameters and stable prewhitening filter.

- Input disturbances $\underline{w}_k^x$ and parameter variations $\underline{w}_k^\delta$ which may be correlated can be equally treated with NKBF.

- The NKBF does not in any way require assumptions as to the stability of the model or the stationarity of the noise processes.

- In cases of additional measurement noise, the NKBF algorithm can be extended by minor modification to a suboptimal state-parameter estimator. In this case, the application in parameter estimation is of particular importance. This aspect will be comprehensively covered in the next section.

I.4 Consideration of measurement noise

In cases where the observations are not free of noise, the optimum solution can not be simply calculated since the a posteriori density $\underline{p}\,(\underline{x}^*_k|\underline{y}^k)$ is no longer Gaussian. Nevertheless, the principle defined in section 3 for noise-free measurements can be used to establish a suboptimal procedure which is presented here, on account of simplicity, for a system in observer form with one input, one output and only measurement noise. The generalization (vector input, vector output, input noise) is not difficult and leads to a formal extension.

Substituting in the observer form

$$\underline{x}_{k+1} = \left[\begin{array}{c|c} 0\ldots0 & \\ \hline \underline{E} & \underline{a} \end{array}\right] \underline{x}_k + \underline{b}\, u_k = \left[\begin{array}{c|c} 0\ldots & 0 \\ \hline \underline{E} & \vdots \\ & 0 \end{array}\right] \underline{x}_k + \underline{a} x^n_k + \underline{b}\, u_k \tag{23}$$

$$y_k = [0\ldots01]\ \underline{x}_k + v_k = x^n_k + v_k \tag{24}$$

the nth component x^n_k of the state vector $\underline{x}_k$ for $x^n_k = \hat{x}^n_k + \tilde{x}^n_k$, where $\hat{x}^n_k$ is the estimate for x^n_k and $\tilde{x}^n_k$ is the estimation error, leads to the extended system with state $\underline{x}^*_k = [\underline{x}^T_k, \underline{a}^T, \underline{b}^T]^T$ (see equations (8) and (9)):

$$\underline{x}^*_{k+1} = \left[\begin{array}{c|c|c} \begin{array}{c|c} 0\ldots & 0 \\ \hline \underline{E} & \vdots\ 0 \end{array} & \hat{x}^n_k \underline{E} & u_k \underline{E} \\ \hline \underline{0} & \underline{E} & \end{array}\right] \underline{x}^*_k + \left[\begin{array}{c} \underline{a}\tilde{x}^n_k \\ \\ \underline{0} \end{array}\right] = \underline{A}^*_k \underline{x}^*_k + \underline{\lambda}^*_k \tag{25}$$

$$y_k = [0\ldots01\,|\,0 \quad \ldots \quad 0]\ \underline{x}^*_k + v_k\ . \tag{26}$$

In the absence of measurement noise x^n_k is identical with y_k and we get with $\tilde{x}^n_k \equiv 0$ the original NKBF. A method of approximation named "Suboptimal Non-Linear Kalman-Filter" - abbreviated: SNKBF - can be defined as follows: Assuming that the error vector $\underline{\lambda}^*_k$ is roughly a white Gaussian noise vector $\underline{w}_k^*$ independent of v_k, then the Kalman-Bucy algorithm - now with consideration of measurement noise - can be applied again. For determination of the filter parameters $\underline{R}_{k+1}$ and $\underline{Q}_k$ (see equations 18 and 21) is suggested:

$$\underline{R}_{k+1} = r = \mathrm{Var}(v_k) = \sigma_v^2 \ , \qquad (27)$$

$$\underline{Q}_k = \begin{bmatrix} \underline{Q}_k^x & \underline{0} \\ \underline{0} & \underline{0} \end{bmatrix} \quad \text{where} \quad \underline{Q}_k^x = q\,\underline{E} = \frac{\sigma_v^2}{4n}\,\underline{E} \ . \qquad (28)$$

In order to understand this determination resulting from empirical examinations /4/, the SNKBF algorithm is examined where r and q are selected differently than in the suggestion above. When $r = 0$, $q > 0$ it follows that $\hat{x}_k^n \equiv y_k$ and the algorithm is identical with the NKBF, whose parameter estimates for adequately large variances pass to the RLQ-estimates which - as $\sigma_v \neq 0$ - are known to be subject to a bias. The other extreme case where $r = \sigma_v^2$, $q = 0$ can be interpreted as a simplified "Extended Kalman-Filter" /9/, /3/, abbreviated EKF. With this method the non-linearities of the system can be developed into a Taylor series which, by neglecting terms of second or higher order, leads to a simplificated estimation. In application as state-parameter-estimator, it differs from the SNKBF in that the matrix $\underline{Q}_k^x$ is essentially more complicated and is dependent on the estimates and their variances. The EKF which approximates the conditional mean works in this application without an essential bias. However, it can lead to divergence effects which must be prevented by introducing a fictitious input noise /9/. Additionally, a more extensive version of the filter equations have to be applied in order to avoid numerical instabilities /9/. In consequence, the effort of calculation associated with the first order EKF applied to the Observer form is approximately four times greater than required by the SNKBF.

Empirical investigations and simlulations on numerous systems have shown that the extra effort associated with the EKF compared with the SNKBF is not justified, i.e. the simple use of $\underline{Q}_k^x = q\underline{E}$ in the

SNKBF leads to a relative good estimate. Further, the suggested value for q has been demonstrated as being a useful compromise through which divergence is avoided and bias is notably diminished compared with the RLQ.

The SNKBF in the simple form of equation (28) can lead to good results, especially in comparison with methods which provide only parameter estimates. This is to be seen in its application to a fourth-order system given in observer form with five unknown parameters. The example is taken from /12/. The parameters are given by

$$\underline{\delta}^T = \left[a_1 \; a_2 \; a_3 \; a_4 \; b \right] = \left[-1.3 \; 0.22 \; 0.832 \; 0.269 \; 1.0 \right]$$

u_k: Pseudo-binary noise signal ($u_k \in \{-1, 1\}$)

v_k: White Gaussian process with $\sigma_v = 0.9$

Fig. 2 shows the impulse response of the system and Fig. 3 the estimation error $\| \underline{\tilde{\delta}}_k \| = \left[\tilde{a}_1^2 + \ldots + \tilde{b}^2 \right]^{1/2}$

Besides the results from 5 known parameter estimation methods which are accurately described in /12/, the mean over 10 realiziations of the estimation error $\overline{\| \underline{\tilde{\delta}}_k \|}$ of the SNKBF is presented. The example shows that not only the striking bias with the RLQ is clearly diminished, but more significantly, it shows the high performance of the SNKBF compared with the other techniques (see cross hatching).

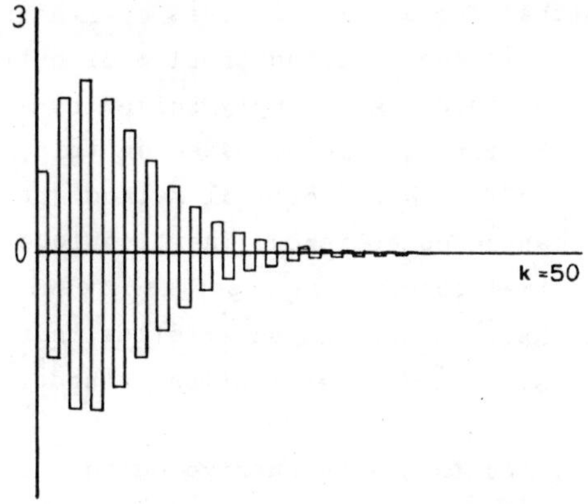

Fig. 2: Impulse response of the system being investigated

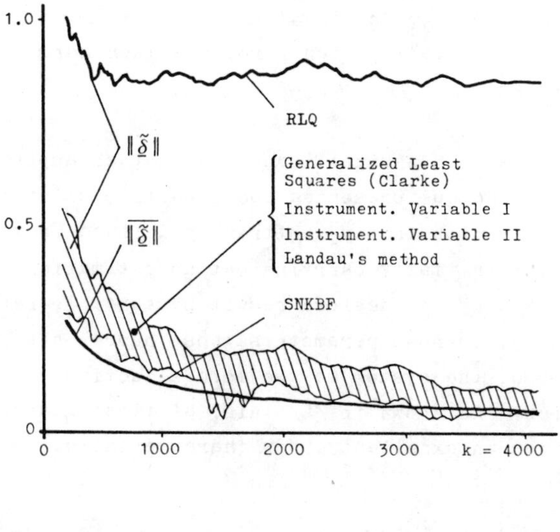

Fig. 3: Comparison of different parameter estimation methods with the SNKBF

I.5 Conclusions

It has been shown that for a certain class of linear systems with noise free measurements,the combined problem of estimating the states and parameters can be relatively simply solved by a special application of the Kalman-Bucy-Filter equations. Based on this concept of solution, and without further effort,a sub-optimal method for systems containing measurement noise can be established. The technique provides for both optimal and sub-optimal cases - as a partial result - parameter estimates that can have better characteristics than those obtained using known classical parameter-estimation methods.

The technique is allied to the "Recursive Method of Least Squares" (RLQ) and the "Extended Kalman-Filter" (EKF) but essentially avoids their disadvantages (bias, divergence).

It is expected that this estimation principle will also lead to new algorithms for the case of unknown noise variances since first, it can be applied in the place of all methods which proceed from modification of the RLQ /1/, /8/, /14/ and second, more free filter parameters are available,which permits a much more flexible processing of disturbances.

The technique also has significance in another application. Using the fundamental technique presented above permits solution of the state-parameter estimation in any arbitrarily defined (physical) state-coordinates,by initially carrying out an estimation in observer form and then obtaining the desired result by suitable retransformation. Simultaneously, unknown parameters whose number can be independent of the order of the system may be more effectively taken into consideration. As shown in /4/ this, at first glance seemingly complex principle, has an excellent error characteristic.

I.6 Appendix

Proof of Theorem 1:

With the substitution

$$\bar{\underline{y}}_j = \underline{y}_j - \underline{g}(j,\underline{y}^{j-1}) ,$$

$$\bar{\underline{y}}^j = \left[\bar{\underline{y}}_o^T \; \dots \; \bar{\underline{y}}_j^T\right]^T$$

equation (12) and (13) can be replaced by

$$\underline{z}_{k+1} = \bar{\underline{\Phi}}(k,\bar{\underline{y}}^k)\underline{z}_k + \bar{\underline{f}}(k,\bar{\underline{y}}^k) + \bar{\underline{\Gamma}}^1(k,\bar{\underline{y}}^k)\,\underline{w}_k$$

$$\bar{\underline{y}}_k = \bar{\underline{C}}(k,\bar{\underline{y}}^{k-1})\underline{z}_k + \bar{\underline{\Gamma}}^2(k,\bar{\underline{y}}^{k-1})\underline{v}_k$$

or in short:

$$\underline{z}_{k+1} = \bar{\underline{\Phi}}_k\underline{z}_k + \bar{\underline{f}}_k + \bar{\underline{\Gamma}}_k^1\underline{w}_k$$

$$\bar{\underline{y}} = \bar{\underline{C}}_k\underline{z}_k + \bar{\underline{\Gamma}}_k^2\underline{v}_k \quad .$$

With the known relationship for the density of two random variables a and b

$$p(a,b) = p(a \mid b)p(b)$$

and

$$\int p(a,b)db = p(a)$$

$p(\underline{z}_o \mid \bar{\underline{y}}_o)$ can be at first expanded to

$$p(\underline{z}_o \mid \bar{\underline{y}}_o) = p(\bar{\underline{y}}_o \mid \underline{z}_o)\frac{p(\underline{z}_o)}{p(\bar{y}_o)} = \frac{p(\bar{\underline{y}}_o \mid \underline{z}_o)p(\underline{z}_o)}{\int p(\bar{\underline{y}}_o \mid \underline{z}_o)p(\underline{z}_o)d\underline{z}_o} \quad .$$

The a posteriori density at time k+1 is given by

$$p(\underline{z}_{k+1}|\bar{\underline{y}}^{k+1}) = \frac{p(\underline{z}_{k+1},\bar{\underline{y}}^{k+1})}{p(\bar{\underline{y}}^{k+1})} = \frac{p(\underline{z}_{k+1},\bar{\underline{y}}^{k+1})}{\int p(\underline{z}_{k+1},\bar{\underline{y}}^{k+1})\,d\underline{z}_{k+1}}$$

$$= \frac{\int p(\underline{z}_{k+1},\underline{z}_k,\underline{y}_{k+1}\ \bar{\underline{y}}^{k})\,d\underline{z}_k}{\int \text{numerator } d\underline{z}_{k+1}}$$

$$= \frac{\int p(\bar{\underline{y}}_{k+1}\ \underline{z}_{k+1},\underline{z}_k,\bar{\underline{y}}^{k})\,p(\underline{z}_{k+1}\ \underline{z}_k,\bar{\underline{y}}^{k})\,p(\underline{z}_k\ \bar{\underline{y}}^{k})\,d\underline{z}_k}{\int \text{numerator } d\underline{z}_{k+1}}$$

with

$$p(\bar{\underline{y}}_{k+1}|\underline{z}_{k+1},\underline{z}_k,\bar{\underline{y}}^{k}) = p(\bar{\underline{C}}_{k+1}\underline{z}_{k+1} + \bar{\underline{\Gamma}}^{2}_{k+1}\underline{v}_{k+1}|\underline{z}_{k+1},\underline{z}_k,\bar{\underline{y}}^{k})$$

and

$$p(\underline{z}_{k+1}|\underline{z}_k,\bar{\underline{y}}^{k}) = p(\bar{\underline{\Phi}}_k\underline{z}_k + \bar{\underline{f}}_k + \bar{\underline{\Gamma}}^{1}\underline{w}_k|\underline{z}_k,\bar{\underline{y}}^{k})$$

Since in general:

$$p(f(y)z|y)\Big|_{y=Y} = p(cz|y) \qquad \text{with } c = f(Y)\ ,$$

all matrices depending of $\bar{\underline{y}}^{k}$ or $\underline{y}^{k}$ can be considered as constants for the density computation. Consequently the problem is formally identical with the state estimation for linear systems and must therefore lead to the Kalman-filter equations (equ. (16) to (22)) by suitable resubstitution.

I.7 References

/1/ Åström-Eykhoff
System Identification - A Survey
Automatica 10/1974

/2/ Brammer, K. / G. Siffling
Kalman-Bucy-Filter
Pretince Hall, 1973

/3/ Brammer, K.
Schätzung von Parametern und Zustandsvariablen linearer Regelstrecken durch nichtlineare Filterung
Regelungstechnik 6/1970

/4/ Bühler, E.
Ein einfaches Optimalfilter zur kombinierten Schätzung von Zustandsgrößen und Parametern linearer Systeme
Dissertation 1979, Technische Universität Berlin

/5/ Eykhoff, P.
System Identification
John Wiley, 1974, New York

/6/ Goodwin, G. / Payne, R.
Dynamic System Identification:
Experiment Design and Data Analysis
Academic Press, New York, 1977

/7/ Hartmann, I.
Lineare Systeme
Springer, Berlin, 1976

/8/ Isermann, R.
Prozeß-Identifikation
Springer, Berlin, 1974

/9/ Jazwinsky, A.H.
Stochastic Prozesses and Filtering Theory
Academic Press, New York, 1970

/10/ Kalman, R.E.
A New Approach to Linear Filtering and Prediction Problems
J.Basic Engg.Bd.82(1966)p.35-45

/11/ Kalman, R.E. / Bucy, R.S.
New Results in Linear Filtering and Prediction Theory
J.Basic Engg.Bd.83(1961)p.95-108

/12/ Landau, I.D.
Unbiased Recursive Identification Using Model Reference Adaptive Techniques
IEEE Trans.on autom. control. No.2/1976

/13/ Landgraf, Ch.
Stochastische lineare Systeme
Brennpunkt Kybernetik, TU Berlin 1976

/14/ Unbehauen/Göhring/Bauer
Parameterschätzverfahren zur Systemidentifikation
Oldenburg Verlag, München, 1974

Part II: State and Parameter Estimation of Linear Systems in Arbitrary State Coordinates

Summary

Most on-line methods for the identification of linear systems are developed for the estimation of coefficients in differential or difference equations. The method presented in this paper is based on a generalized model description which includes arbitrary, unknown parameters; e.g. mass, temperature, resistance. The method can be used for parameter estimation as well as for adaptive state estimation in arbitrary state coordinates.

II.1 Introduction

Many processes can be adequately described by a linear, dynamical model. However, it is not always possible to measure or calculate the model parameters and states with the necessary accuracy. In this case, on-line identification may be helpful.

For most identification concepts, the coefficients $\underline{\delta} = [\delta_1 \cdots \delta_q]^T$ of differential or difference equations are estimated. This leads to relatively simple algorithms, but in most cases the number of coefficients $\underline{\delta}$ is greater than the number of really unknown parameters $\underline{\nu} = [\nu_1 \cdots \nu_m]^T$ (e.g. mass, temperature, resistance) upon which, in general, $\underline{\delta}$ depends in a nonlinear manner: $\underline{\delta} = \underline{\psi}(\underline{\nu})$. This implies a interdependence of certain elements of $\underline{\delta}$, which, by appropriate incorporation, may lead to a better estimation than is possible when all δ_i are assumed to be unknown and freely definable.

The concept presented in this paper emphasizes this fact. It is, in addition to parameter identification, applicable to combined estimation of states and parameters; in short: state-parameter-estimation. In this case, the state coordinates may be arbitrarily given. This means that the canonical model structures demanded by many concepts need not be required here.

The main aspect of this on-line-applicable method is the separation of the identification procedure into two steps. The first step can be traced back to well-known methods and will not be considered comprehensively here. It contains the recursive estimation of the vector $\underline{\delta}$ and, if need be, of the state vector $\underline{x}$, where it is assumed that all the elements of $\underline{\delta}$ may be freely definable (independent), and have Gaussian a priori distribution. Furthermore, if a state estimation is desired, the model is set forth for simplification in a particular canonical form: the observer form.
For a detailed exposition of this point and the derivation of appropriate approaches see /4/.

Based on the results from step one, information about the function $\underline{\delta} = \underline{\psi}\ (\underline{\nu})$ and about a probable given, a priori distribution of $\underline{\nu}$ is utilized in a second, statical step, to obtain estimates $\hat{\underline{\nu}}$, and if dim $(\underline{\delta})$ > dim $(\underline{\nu})$, to improve the estimates of the state $\underline{x}$. If necessary, these estimates may subsequently be transformed back to the original state coordinates.

Although this concept looks complicated at first, separation of the identification process into two steps introduces essential advantages, especially in those cases where the number m=dim$(\underline{\nu})$ of the originally unknown parameters is relatively small, and if not all states must be transformed back, which is the case in many applications.

Contrary to well-known, on-line-applicable methods, e.g./3/, the two-step concept makes possible the calculation of the non-linearities in step 2 outside the recursion loop. This implies that errors which arise from simple approximations cannot carry over from one sample to the other. For the same reason, relatively rough approximations may be employed in step two, with a great saving of calculation time without loss of good error estimability.

II.2 Statement of the problem

Start with the following linear system of order n, which is described, with respect to the use of the digital computer, in the discrete representation:

$$\left.\begin{aligned} \underline{z}_{k+1} &= \underline{\Phi}(\underline{\nu})\,\underline{z}_k + \underline{H}(\underline{\nu})\,\underline{u}_k + \underline{w}_k^1 \\ \underline{y}_k &= \underline{C}(\underline{\nu})\,\underline{z}_k + \underline{w}_k^2 \end{aligned}\right\} \quad (2.1)$$

$\underline{z}_k$ state vector with the dimension n

$\underline{u}_k$ input vector with the dimension r

$\underline{y}_k$ output vector with the dimension s

$$\left.\begin{aligned} &\underline{\Phi} \quad n \times n \\ &\underline{H} \quad n \times r \\ &\underline{C} \quad s \times n \end{aligned}\right\} \quad \text{matrices}$$

$\underline{\nu}$ unknown parameter vector of the dimension m.

$\underline{w}_k^1$ and $\underline{w}_k^2$ represent disturbances, which are assumed in the following as white, independent Gaussian processes.

Let $\underline{u}_k$ and $\underline{y}_k$ be measurable (assumed to be known). The elements of the matrices $\underline{\Phi}$, $\underline{H}$, $\underline{C}$ may depend, in a non-linear manner, on the parameters summarized in the vector $\underline{\nu}$.

This general statement of the problem may result if the model (2.1) is taken from the continuous system by discretization. This is shown by a simple example:

From the continuous system

$$\begin{aligned} \dot{z} &= \nu z + u \\ y &= z\ . \end{aligned}$$

follows the discretized system

$$\begin{aligned} z_{k+1} &= \delta_1 z_k + \delta_2 u_k \\ y_k &= z_k \end{aligned}$$

with $\delta_1 = e^{\gamma T}$, $\delta_2 = \frac{1}{\gamma} \langle e^{(\gamma T)} - 1 \rangle$, T = sampling time .

The problem of combining estimation of states and parameters can be solved suggestively only if the system is observable. This condition must be fulfilled for all values of $\underline{\gamma}$; this means that one has to ensure that:

$$\underline{\gamma} \in \Sigma$$

$$\text{with } \Sigma = \{ \underline{\gamma} \,|\, (\underline{\Phi}(\underline{\gamma}) , \underline{C}(\underline{\gamma})) \text{ observable} \} . \tag{2.2}$$

Based on this assumption, there must be for every $\underline{\gamma} \in \Sigma$ a system in observer form which can be obtained from (2.1) through a linear transformation, which can be described as follows (one input, one output):

$$\begin{aligned} \underline{x}_{k+1} &= \underline{A}\,\underline{x}_k + \underline{b}\,u_k = \left[\begin{array}{c|c} 0 \; \dots \; 0 & \\ \underline{E} & \underline{a} \end{array} \right] \underline{x}_k + \underline{b}\,u_k \\ y_k &= \underline{e}_n^T \, \underline{x}_k \qquad\quad = [0 \dots 0 1] \quad \underline{x}_k . \end{aligned} \tag{2.3}$$

This canonical structure belongs, as a special case, to a particular class of systems which is of fundamental importance for the state-parameter estimation presented in /4/.

The elements a_i, b_i, from $\underline{a}$, $\underline{b}$, which appear directly in the z-transfer function

$$G(z) = \frac{b_n z^{n-1} + \dots + b_2 z + b_1}{z^n - a_n z^{n-1} - \dots - a_2 z - a_1}$$

are summarized in the vector

$$\underline{\delta} := \begin{bmatrix} \underline{a} \\ \underline{b} \end{bmatrix} \tag{2.4}$$

of dimension q (q = 2n for single-input, single-output systems).

The relation between $\underline{\gamma}$ and $\underline{\delta}$, which plays an important role in the realization of an identifier, is defined by the following equation (single input, single output):

$$\left\{ \underline{A}(\underline{\delta}), \ \underline{b}(\underline{\delta}), \ \underline{e}_n^T \right\} = \left\{ \underline{T}(\underline{\gamma}) \underline{\Phi}(\underline{\gamma}) \underline{T}^{-1}(\underline{\gamma}), \underline{T}(\underline{\gamma}) \underline{h}(\underline{\gamma}), \ \underline{T}^{-1}(\underline{\gamma}) \underline{c}(\underline{\gamma}) \right\} \quad (2.5)$$

with $\underline{T}(\underline{\gamma})$ as a regular transformation matrix for the state $\underline{x}_k = T(\underline{\gamma}) \ \underline{z}_k$. This equation defines the map

$$\underline{\gamma} \longmapsto \underline{\phi}(\underline{\gamma}) \qquad\qquad \text{for } \underline{\gamma} \in \Sigma \ . \qquad\qquad (2.6)$$

The domain of this map is given by the equation (2.2). The range is named by Ω :

$$\Omega := \left\{ \ \underline{\delta} \ \mid \ \underline{\delta} = \underline{\phi}(\underline{\gamma}) \ , \quad \underline{\gamma} \in \Sigma \ \right\} . \qquad (2.7)$$

Besides observability, the existence of a consistence parameter estimation is required:

$$\hat{\underline{\gamma}}_k = t_k^\gamma(\underline{y}^k) \ , \ \text{whereby} \ \hat{\underline{\gamma}}_k \longrightarrow \underline{\gamma} \ \text{for } k \longrightarrow \infty$$

$$\text{with} \qquad \underline{y}^k := \left[\underline{y}_o \ \cdots \ \underline{y}_k \right]^T.$$

This means that $\hat{\underline{\gamma}}_k$ has to converge to the true parameter $\underline{\gamma}$ in a stochastic sense /6/. The conditions leading to a consistence estimation $\hat{\underline{\delta}}_k = t_k^\delta(\underline{y}^k)$ are well-known in the literature, /2/, /12/. Supposing that these conditions have been fulfilled, then there must also be a consistence estimation $\hat{\underline{\gamma}} = t_k^\gamma(\underline{y}^k)$ if the map $\underline{\phi}$ exhibits some particular characteristics of of continuity (always fulfilled in practice) and if the map $\underline{\phi}$ is invertible. This should be taken as true for the following. For a more detailed treatment of identifiability see /5/, and /4/.

In many cases invertibility must be forced by introducing a new definition of $\underline{\gamma}$ or by a constriction of Σ to $\tilde{\Sigma}$, as demonstrated by the following example:

Given the system of order two (oscillator)

$$\dot{\underline{\xi}}(t) = \begin{bmatrix} 0 & 1 \\ \omega^2 & 0 \end{bmatrix} \underline{\xi}(t), \qquad \underline{y}(t) = \begin{bmatrix} 1 & 0 \end{bmatrix} \underline{\xi}(t) \quad . \qquad (2.8)$$

By discretization, with T as sampling time, we obtain

$$\begin{aligned} \underline{z}_{k+1} &= \underline{\Phi}(T)\underline{z}_k \\ y_k &= \begin{bmatrix} 1 & 0 \end{bmatrix} \underline{z}_k \end{aligned} \quad \text{with } \underline{\Phi}(T) = \begin{bmatrix} \cos\omega T & \frac{1}{\omega}\sin\omega T \\ -\omega\sin\omega T & \cos\omega T \end{bmatrix} \qquad (2.9)$$

Transformation to the observer form leads to

$$\begin{aligned} \underline{x}_{k+1} &= \underline{A}\,\underline{x}_k \\ y_k &= \begin{bmatrix} 0 & 1 \end{bmatrix} \underline{x}_k \end{aligned} \quad \text{with } \underline{A} = \begin{bmatrix} 0 & a_1 \\ 1 & a_2 \end{bmatrix} \qquad \begin{aligned} a_1 &= -1 \\ a_2 &= 2\cos\omega T \end{aligned} \qquad (2.10)$$

From the condition of observability /7/

$$\text{Rank} \begin{bmatrix} \underline{c}^T \\ \underline{c}^T \underline{\Phi} \end{bmatrix} = \text{Rank} \begin{bmatrix} 1 & 0 \\ \cos\omega T & \frac{1}{\omega}\sin\omega T \end{bmatrix} \stackrel{!}{=} 2$$

follows the observability for all $\omega \neq i\,\frac{\pi}{T}$, $i = 1,2,3\ldots$. Thus the domain and range of $\underline{\psi}$ are given by (see Figure 1):

$$\Sigma = \{\, \omega \in \mathbb{R} \mid \omega \neq i\,\frac{\pi}{T}\,, \quad i = 1,2,3\ldots \}$$

$$\Omega = \{\, \underline{a} \in \mathbb{R}^2 \mid -2 < a_2 \leqslant 2\,, \quad a_1 = -1 \,\}$$

$$\underline{\delta} = \underline{a} = \underline{\psi}(\underline{\gamma}) = \underline{\psi}(\omega) = \begin{bmatrix} -1 \\ 2\cos\omega T \end{bmatrix} .$$

As can be seen, the map is not invertible. This can be rectified by constricting the domain to $\tilde{\Sigma} \subset \Sigma$ with

$$\tilde{\Sigma} = \{\, \omega \mid 0 \leqslant \omega < \omega_{max} = \pi/T \,\} . \qquad (2.11)$$

This is exactly the condition of Shannon's sampling theorem, which states that a harmonic signal can be reconstructed only if the sampling time is less than half the period of the signal. This means that an upper bound for all ω values has to be known in advance. This bound determines the maximal sampling time within which a unique identification of the parameter ω is just possible.

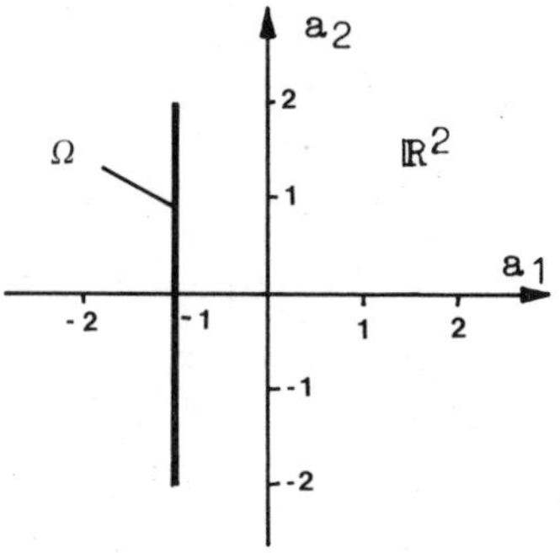

Figure 1: Range of the map ψ

In the following sections, $\underline{\nu}$ is assumed to be a random variable independent of k and with a known a priori distribution $p(\underline{\nu})$.

Since states and parameters can be comprised in an augmented vector

$$\underline{z}_k^* = \begin{bmatrix} \underline{z}_k \\ \underline{\nu} \end{bmatrix} \qquad (2.12)$$

of a dynamical system

$$\left.\begin{aligned} \underline{z}_k^* &= \underline{f}(k,\ \underline{z}_k^*,\ \underline{u}_k,\ \underline{w}_k^1) \\ \underline{y}_k &= \underline{g}(k,\ \underline{z}_k^*,\ \underline{u}_k,\ \underline{w}_k^2) \end{aligned}\right\} \qquad (2.13)$$

the problem of simultaneous estimation of states and parameters can be interpreted as a non-linear filtering problem (state filtering)/3/, the theoretical solution of which is the calculation of the a posteriori distribution $p(\underline{z}_k^* | \underline{y}^k)$ of $\underline{z}_k^*$ given the observations $\underline{y}^k$. In most cases, instead of the distribution $p(\underline{z}_k^* | \underline{y}^k)$, only an estimate $\hat{\underline{z}}_k^*$ is of interest (e.g. maximum a posteriori estimation, conditional mean) because the numerical requirements are not so large.

In addition to the augmented state vector $\underline{z}_k^*$ as defined by (2.12), the corresponding vector

$$\underline{x}_k^* = \begin{bmatrix} \underline{x}_k \\ \\ \underline{\delta}_k \end{bmatrix} \tag{2.14}$$

as an augmented state vector of the system in observer form is of interest. It is related to $\underline{z}_k^*$ as follows

$$\underline{x}_k^* = \begin{bmatrix} \underline{x}_k \\ \\ \underline{\delta}_k \end{bmatrix} = \begin{bmatrix} \underline{T}(\underline{\nu}_k)\ \underline{z}_k \\ \\ \underline{\psi}(\underline{\nu}_k) \end{bmatrix} = \underline{\psi}^*\ (\underline{z}_k^*)\ . \tag{2.15}$$

Since $\underline{T}(\underline{\nu})$ is a non-singular matrix for $\underline{\nu} \in \Sigma$, invertibility of $\underline{\psi}$ implies invertibility of $\underline{\psi}^*$ and vice versa .

II.3 Calculation of the a posteriori distributions

Before estimation functions for $\underline{\nu}$ and $\underline{z}_k^*$ are derived in sections 4 and 5, it will be shown in this section how to calculate the distributions $p(\underline{\nu}\mid \underline{y}^k)$ and $p(\underline{z}_k,\underline{\nu}|\underline{y}^k) = p(\underline{z}_k^*\mid \underline{y}^k)$ from the distributions $p(\underline{\delta}\mid \underline{y}^k)$ and $p(\underline{x}_k^*\mid \underline{y}^k)$, which are assumed to be given as a result of the first step.

At first, Bayes' rule will be modified. For the variable x, given the observation y, we can state:

$$\begin{aligned} p(x|y) &= \frac{1}{p(y)}\, p(y|x)p(x) \\ &= \frac{1}{\bar{p}(y)}\, p(y|x)\bar{p}(x) \cdot g(x) = \bar{p}(x|y)\ g(x) \end{aligned} \tag{3.1}$$

with

$$\left.\begin{aligned} \bar{p}(x|y) &= \frac{1}{\bar{p}(y)}\, p(y|x)\bar{p}(x)\ , \\ g(x) &= \frac{\bar{p}(y)}{p(y)}\,\frac{p(x)}{\bar{p}(x)} = c'\,\frac{p(x)}{\bar{p}(x)}\ . \end{aligned}\right\} \tag{3.2}$$

$\bar{p}(x)$ is an arbitrary distribution; but let the quotient $p(x)/\bar{p}(x)$ exist. Further, let $\bar{p}(y)$ relating to $\bar{p}(x)$ be not equal to zero for the observation y.

The practical significance of this procedure is that the a posteriori distribution $p(x|y)$ emerging from $p(x)$ can be reverted to the a posteriori distribution $\bar{p}(x|y)$, which relates to a hypothetical a priori distribution $\bar{p}(x)$. $\bar{p}(x)$ has to be chosen in such a way that the calculation of $\bar{p}(x|y)$ will be simple - e.g.: $\bar{p}(x)$ stated as a Gaussian distribution. If in this case $\bar{p}(x) = N(0\, ,\, \frac{1}{\epsilon})$ [1], for a sufficiently small ϵ $\bar{p}(x)$ may be assumed to be constant over the interesting region. Hence it holds

$$g(x) \approx c\, p(x) \qquad ^{2)} \tag{3.3}$$

$$p(x|y) \approx c\, \bar{p}(x|y;\epsilon)p(x) \quad . \tag{3.4}$$

After these introductory lines, and with the well-known relations

$$p_{c|b}(c|b) = p_{c|a}(c|f(b)) \quad , \quad ^{3)} \tag{3.5}$$

$$p_a(f(b)) = p_b(f^{-1}(a)) \left|\det \frac{\partial f^{-1}(a)}{\partial a}\right| \tag{3.6}$$

with $a = f(b)$ as an invertible function

at first the combined estimation of states and parameters will be considered and a formula for $p(\underline{z}_k^*|\underline{y}^k)$ will be derived.

Next to the unknown distribution of $p(\underline{z}_k^*|\underline{y}^k)$ holds:

$$p(\underline{z}_k^*|y^k) = c\, \frac{p_{y|z^*}(\underline{y}^k|\underline{z}_k^*)p(\underline{z}_k^*)}{p_{y|x^*}(\underline{y}^k|\underline{\psi}^*(\underline{z}_k^*))p(z_k^*)} \quad . \tag{3.7}$$

---

1) This unconventional notation is used here for the sake of clarity.

2) In this equation and in the following, c represents a scaling factor independent of x.

3) If necessary, the notation $p(a|b)$ will be replaced more precisely with $p_{a|b}(\cdot|\cdot)$.

$p(\underline{z}_k^*)$ can be remodeled as

$$p(\underline{z}_k^*) = p(\underline{z}_k, \underline{\gamma}) = p(\underline{z}_k|\underline{\gamma})p(\underline{\gamma})$$

which, with the linear state transformation $\underline{x}_k = \underline{T}\ \underline{z}_k$ leads to

$$\begin{aligned} p(\underline{z}_k^*) &= p_{z|\gamma}(\underline{z}_k|\underline{\gamma})\ p(\underline{\gamma}) \\ &= p_{x|\gamma}(\underline{T}\ \underline{z}_k|\underline{\gamma})\ |\det(\underline{T})|\, p(\underline{\gamma}) \\ &= p_{x|\delta}(\underline{T}\ z_k|\underline{\psi}(\underline{\gamma}))\ |\det(\underline{T})|\, p(\underline{\gamma})\ . \end{aligned}$$

With respect to equation (3.1), it follows that

$$\begin{aligned} p(\underline{z}_k^*) &= p_{x|\delta}(\underline{T}(\underline{\gamma})\underline{z}_k\,|\,\underline{\psi}(\underline{\gamma}))\ |\det\underline{T}(\underline{\gamma})|\ p(\underline{\gamma}) \\ &= p_{x|\delta}(\underline{T}(\underline{\gamma})\underline{z}_k\,|\,\underline{\psi}(\underline{\gamma}))\bar{p}_\delta(\underline{\psi}(\underline{\gamma}))\ |\det\underline{T}(\underline{\gamma})|\ \frac{p(\underline{\gamma})}{\bar{p}_\delta(\underline{\psi}(\underline{\gamma}))} \\ &= \bar{p}_{x,\delta}(\underline{T}(\underline{\gamma})\underline{z}_k,\ \underline{\psi}(\underline{\gamma}))\ |\det(\underline{T})|\ \frac{p(\underline{\gamma})}{\bar{p}_\delta(\underline{\psi}(\underline{\gamma}))} \\ &= \bar{p}_{\underline{x}^*}(\underline{\psi}^*(\underline{z}_k^*))\ g^*(\underline{\gamma})\ . \end{aligned}$$

With equation (3.7), we finally get:

$$p(\underline{z}_k^*|\underline{Y}^k) = c\ p_{Y|x^*}(\underline{Y}^k|\ \underline{\psi}^*(\underline{z}_k^*))\bar{p}_{x^*}(\ \underline{\psi}^*(\underline{z}_k^*))\ g^*(\underline{\gamma})$$

or

$$\boxed{p(\underline{z}_k^*|\underline{Y}^k) = c\ \bar{p}_{x^*|Y}(\ \underline{\psi}^*(\underline{z}_k^*)\ |\ \underline{Y}^k)\ g^*(\underline{\gamma})} \qquad (3.8)$$

with $\underline{\psi}^*(\underline{z}_k^*) = \begin{bmatrix} T(\underline{\gamma})z_k \\ \\ \underline{\psi}(\underline{\gamma}) \end{bmatrix}$, $g^*(\underline{\gamma}) = |\det(\underline{T}(\underline{\gamma})|\ \dfrac{p_\gamma(\underline{\gamma})}{\bar{p}_\delta(\underline{\psi}(\underline{\gamma}))}$

In this equation,

$c = c(\underline{y}^k)$: Scaling factor which is unimportant for maximization (maximum a posteriori estimation)

$\bar{p}_{x^*|y}(\underline{\psi}^*(\underline{z}_k^*)|\underline{y}^k)$: a posteriori distribution of the state-parameter vector in observer form, based on the hypothetical a priori distribution.

$g^*(\underline{\gamma})$: Correction term.

To be able to calculate $\bar{p}(\underline{x}^*|\,\underline{y}^k)$ simply, that is, by the method presented in /4/, one should choose $\bar{p}(\underline{\delta}) = N(\underline{O}, \frac{1}{\epsilon}\,\underline{E})$.

Since $\underline{T}(\underline{\gamma})$ is a non-singular transformation for all $\underline{\gamma} \in \Sigma$ then $\det(\underline{T}(\underline{\gamma})) \neq 0$. $g^*(\underline{\gamma})$ is computable in advance, and may be stored with sufficient accuracy by a suitable approximation. From equation (3.8), the calculation of the conditional density for states and parameters in an arbitrary state representation is led back to the calculation of the conditional density of the corresponding values of the observer form by the hypothesis that the parameters are distributed with $\bar{p}(\underline{\delta})$.

These considerations are valid for any distribution. To get simple conditions with respect to an estimation function (see sections 4 and 5), in the following, $\bar{p}(\underline{\delta}|\underline{y}^k)$ or $\bar{p}(\underline{x}_k, \underline{\delta}\,|\underline{y}^k)$ are assumed to be Gaussian; or it is assumed that the first two central moments of this density can be calculated by well-known methods.

Since $\underline{z}_k$ - contrary to $\underline{\gamma}$ - is dependent on the time index k, it will be scarcely possible to subsequently incorporate a given a priori density of the initial state $\underline{z}_o$ by a correction density, as may be done with $p(\underline{\gamma})$. Hence $p(\underline{z}_o)$ must be converted from $p(\underline{x}_o)$ and included in the calculation of $\bar{p}(\underline{x}_k^*|\underline{y}^k)$ as an initial density. This will lead to simple results only if $\underline{x}_o$ has a Gaussian distribution $\underline{x}_o \sim N(\bar{\underline{x}}_o, \underline{P}_o^x)$. This restriction is not essential, since mainly two cases, which do not lead to difficulties, need to be considered for application. In the first case, at the beginning of the identification or state-parameter estimation, the initial state $\underline{z}_o = \underline{0}$ - and hence $\underline{x}_o = \underline{0}$ is known (e.g. at the launching of a missile), so $\underline{x}_o \sim N(\underline{0}, \underline{0})$ is to be stated.

This additional information "$\underline{x} = \underline{0}$" naturally means greater accuracy, especially in the initial phase of the estimation. That is, the number of observations necessary to determine the unknown parameters in a linear, noise-free nth-order system will be reduced by n.

In the second case there is no a priori information about the course of the state trajectory until the beginning of the estimation. Now the assumption $\underline{x}_o \sim N(\underline{0}, \frac{1}{\epsilon}\underline{E})$, with sufficiently small ϵ can be made. This means no a priori information about $\underline{x}_o$ is included in the estimation (e.g. identification of a missile with an unknown starting point and unknown starting velocity).

$p(\underline{z}_k^*|\underline{y}^k)$ will seldom be calculated by on-line methods. However, knowledge of the density may be of interest in problems from test theory, as well as for comparisons having approximate solutions.

In accordance with the derivation of equation (3.8) we obtain a formula for the a posteriori density of the parameters as follows:

$$
\begin{aligned}
p(\underline{\nu}|\underline{y}^k) &= c\, p_{y|\nu}(\underline{y}^k|\underline{\nu})p(\underline{\nu}) \\
&= c\, p_{y|\delta}(\underline{y}^k|\underline{\psi}(\underline{\nu}))p(\underline{\nu}) \\
&= c\, p_{y|\delta}(\underline{y}^k|\underline{\psi}(\underline{\nu}))\bar{p}_\delta(\underline{\psi}(\underline{\nu}))\, \frac{p(\underline{\nu})}{\bar{p}_\delta(\underline{\psi}(\underline{\nu}))}
\end{aligned}
$$

or with $\bar{p}(\underline{\delta}|\underline{y}^k) = c\, p(\underline{y}^k|\underline{\delta})\bar{p}(\underline{\delta})$:

$$\boxed{p(\underline{\nu}|\underline{y}^k) = c\, \bar{p}_{\delta|y}(\underline{\psi}(\underline{\nu})\,|\,\underline{y}^k)g(\underline{\nu})} \tag{3.9}$$

and in particular for $\bar{p}(\underline{\delta}) = N(\underline{0}, \frac{1}{\epsilon}\underline{E})$:

$$p(\underline{\nu}|\underline{y}^k) \approx c\, \bar{p}_{\delta|y}(\underline{\psi}(\underline{\nu})|\underline{y}^k\,;\,\epsilon)p(\underline{\nu}) \,. \tag{3.9a}$$

With respect to the estimation functions described in sections 4 and 5, four further densities are relevant and are presented in the following. Their derivation is given in the appendix.

$$p(\underline{x}_k, \underline{\nu} | \underline{y}^k) \quad = \quad c\ p_{x,\delta\ |\ y}(\underline{x}_k, \underline{\psi}(\underline{\nu})\ |\ \underline{y}^k) g(\underline{\nu}) \tag{3.10}$$

If ϵ is sufficiently small, then:

$$p(\underline{y}^k | \underline{z}_k^*) \quad \approx \quad c\ \bar{p}_{x^*|\ y}(\underline{\psi}^*(\underline{z}_k^*)\ |\ \underline{y}^k\ ;\ \epsilon) \tag{3.11}$$

$$p(\underline{y}^k | \underline{x}_k^*) \quad \approx \quad c\ \bar{p}_{x^*|\ y}(\underline{x}_k^* | \underline{y}^k;\ \epsilon\ ) \tag{3.12}$$

$$p(\underline{y}^k | \underline{\nu}) \quad \approx \quad c\ \bar{p}_{\delta\,|y}(\underline{\psi}(\underline{\nu}) | \underline{y}^k;\ \epsilon\ ) \tag{3.13}$$

Figure 2 demonstrates with a simple example how the density $p(\underline{\nu}|\underline{y}^k)$ emerges from $\bar{p}(\underline{\delta}|\underline{y}^k)$ when $\underline{\delta}=\underline{\psi}(\underline{\nu})$ is taken into account. $\underline{\delta} \in \Omega \subset \mathbb{R}^2$ and $\nu \in \mathbb{R}^1$ are time-independent random variables which are bound by a unique relation $\underline{\psi} : \mathbb{R} \to \Omega$. $\bar{p}(\underline{\delta}|\underline{y}^k)$ may be interpreted as the a posteriori density of $\underline{\delta}$ which doesn't include the real connections between the elements of the parameter vector $\underline{\delta}$ but which is developed under the hypothesis of a Gaussian a priori distribution. In general, $\bar{p}(\underline{\delta}|\underline{y}^k)$ is expanded around both coordinates. The actual a posteriori density $p(\underline{\delta}\ |\ \underline{y}^k)$, incorporating $\underline{\psi}$ and $p(\underline{\nu})$, is not represented in figure 2, for reasons of clarity. This density is given by

$$p(\underline{\delta}|\underline{y}^k) \quad = \quad \bar{p}(\underline{\delta}|\underline{y}^k) g(\underline{\delta}) \quad \approx \quad c\ \bar{p}(\underline{\delta}|\underline{y}^k) p(\underline{\delta}) \tag{3.14}$$

with $p(\underline{\delta}) = 0$ for $\underline{\delta} \notin \Omega$. Hence $p(\underline{\delta}|\underline{y}^k)$ - like $p(\underline{\delta})$ - is a singular density, concentrated to Ω.

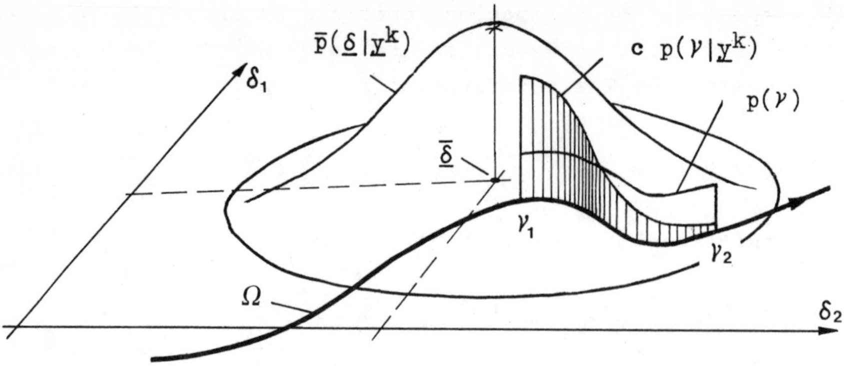

Figure 2: Derivation of $c\ p(\underline{\nu}|\underline{y}^k)$ from $\bar{p}(\underline{\delta}|\underline{y}^k)$

This example shows how the region of all possible values for $\underline{\delta} \in \mathbb{R}^q$ can be essentially restricted by incorporating $\underline{\psi}$ and indeed all the more the greater the difference $\dim(\underline{\delta}) - \dim(\underline{\nu})$ will be. Further, concentration is achieved through the weighting of a correction function $g(\nu)$ or an a priori density $p(\nu)$, which, in Figure 2, is assumed to be constant over the interval (ν_1, ν_2). The set Ω in this example, a line - is in general an arbitrary complicated subset of the parameter space $\mathbb{R}^q$.

It is clear that the inclusion of $\underline{\psi}$ and $p(\underline{\nu})$ influences not only the quality of the parameter estimation, but may also essentially decrease the variance of the state estimation (see Equation (3.8)). This point will be considered in more detail in section 5.

The above-mentioned two-step method can now be represented more exactly. Figure 3 shows the identification concept of the state-parameter estimation problem. The parameter estimation may be treated in a similar manner.

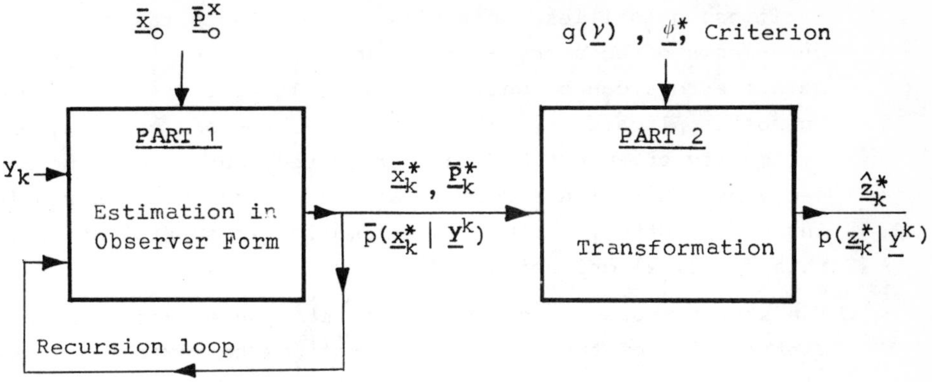

Figure 3: The two-step concept

The first, recursive part of the method leads to the a posteriori density $\bar{p}(\underline{x}_k^* | \underline{y}^k) = \bar{p}(\underline{x}_k, \underline{\delta} | \underline{y}^k)$ of states and parameters in the observer form. Under certain conditions this density is Gaussian, or may be approximated as Gaussian /4/. In the latter case, the a priori density of $p(\underline{\delta})$ of the parameters will be replaced by a better conditioned density $\bar{p}(\underline{\delta})$. A suggestive choice is $\bar{p}(\underline{\delta}) = N(\underline{0}, \frac{1}{\epsilon}\underline{E})$.

In the second part of the identification which is considered in this paper, information about $\underline{\psi}$ and $p(\underline{\nu})$, as well as $g(\underline{\nu})$ is used to obtain the distribution of $p(\underline{z}_k^* | \underline{y}^k)$ from $\bar{p}(\underline{x}_k^* | \underline{y}^k) = N(\bar{\underline{x}}_k^*, \bar{\underline{P}}_k^*)$ or to get estimates of $\underline{z}_k$ from $\bar{\underline{x}}_k^*$, $\bar{\underline{P}}_k^*$ by taking a loss function into account (see sections 4 and 5).

This separation of the identification as shown in Figure 3 leads to the following practical advantages:

- Application of simple, known methods in the first step.
- No feedback from the second step to the recursive calculation of $\bar{\underline{x}}_k^*$, $\bar{\underline{P}}_k^*$ or $\bar{p}(\underline{x}_k^* | \underline{y}^k)$. The numerical processing of the nonlinear function $\underline{\psi}$ is done externally from the recursion loop, which prevents the transmission of errors from one of these sampling steps to the other.

- By appropriate approximation (see section 4) of the non-linearities, quantities can easily be matched to the given requirements of accuracy, whereby now, as before, the approximation errors can be well estimated. It must be noted that in most applications, not consistency, absence of bias, or optimality of an estimation is required, but rather the fast decay of estimation errors to a more or less greater remaining error. Often, a rough approximation is sufficient for this practical criterion.
- The static character of the second step makes possible the distribution of the numerical operations to several samples without essential loss of accuracy. This leads to an high decrease in processing time.
- The non-linearities need not be assumed to be differentiable or linearizable.
- The technique works globally, which means that arbitrary, initial errors can be permitted.

II.4 Parameter estimation

In most applications, instead of densities, estimates are required. The following consideration leads at first to the derivation of parameter-estimation functions. Appropriate functions are the Maximum a priori Estimation:

$$\hat{\underline{\gamma}}_k^{MAP} = \arg \max_{\underline{\gamma} \in \Sigma} p(\underline{\gamma} | \underline{y}^k)$$ 1)

and the Maximum-likelihood Estimation

$$\hat{\underline{\gamma}}_k^{ML} = \arg \max_{\underline{\gamma} \in \Sigma} p(\underline{y}^k | \underline{\gamma})$$

$\hat{\underline{\gamma}}_k^{MAP}$ is obtained by maximizing $p(\underline{\gamma}|\underline{y}^k)$ in equation (3.9) and (3.9a) respectively. With the assumption $\bar{p}(\underline{\delta}|\underline{y}^k) = N(\bar{\underline{\delta}}_k, \bar{\underline{P}}_k^{\delta})$

---

1) arg f(x) := x

follows

$$\hat{\underline{\nu}}_k^{MAP} = \arg\max_{\underline{\nu}\in\Sigma} p(\underline{\nu}|\underline{y}^k) = \arg\max_{\underline{\nu}\in\Sigma} p_{\delta|y}(\underline{\psi}(\underline{\nu})|\underline{y}_k) g(\underline{\nu})$$

$$= \arg\max_{\underline{\nu}\in\Sigma} \left\{ \ln p_{\delta|y}(\underline{\psi}(\underline{\nu}) \mid \underline{y}^k) + \ln g(\underline{\nu}) \right\}$$

$$= \arg\min_{\underline{\nu}\in\Sigma} \left\{ -\left\langle \ln c - \frac{1}{2}\left\{ (\underline{\psi}(\underline{\nu}) - \bar{\underline{\delta}}_k)^T (\bar{\underline{P}}_k^{\delta})^{-1} (\underline{\psi}(\underline{\nu}) - \bar{\underline{\delta}}_k) \right\} + \ln g(\underline{\nu}) \right\rangle \right\}$$

or

$$\hat{\underline{\nu}}_k^{MAP} = \arg\min_{\underline{\nu}\in\Sigma} J_k(\underline{\nu}) \tag{4.1}$$

with $J_k = (\underline{\psi}(\underline{\nu}) - \bar{\underline{\delta}}_k)^T (\bar{\underline{P}}_k^{\delta})^{-1} (\underline{\psi}(\underline{\nu}) - \bar{\underline{\delta}}_k) - \ln g(\underline{\nu})$.

Starting from equation(3.13) we obtain accordingly

$$\hat{\underline{\nu}}_k^{ML} = \arg\min_{\nu\in\Sigma} \; (\underline{\psi}(\underline{\nu}) - \bar{\underline{\delta}}_k)^T (\bar{\underline{P}}_k^{\delta})^{-1} (\underline{\psi}(\underline{\nu}) - \bar{\underline{\delta}}_k) \; . \tag{4.2}$$

With the assumption that $J_k(\underline{\nu})$ is twice differentiable to $\underline{\nu}$, the conditions for a local minimum are given by

$$\frac{\partial J_k(\underline{\nu})}{\partial \underline{\nu}} = 2 \frac{\partial \underline{\psi}^T(\underline{\nu})}{\partial \underline{\nu}} (\bar{\underline{P}}_k^{\delta})^{-1} (\underline{\psi}(\underline{\nu}) - \bar{\underline{\delta}}_k) + \frac{\partial}{\partial \nu} \ln g(\underline{\nu}) = \underline{0} \tag{4.3}$$

and

$$\frac{\partial^2 J_k(\underline{\nu})}{\partial \underline{\nu}^T \partial \underline{\nu}} \quad \text{positive definite .}$$

The numerical solution could be obtained by many known approaches, such as gradient methods. However, to avoid suppositions about differentiability and to make allowance for the particular form of the loss function $J(\underline{\nu})$, and for practical requirements to an on-line identification, the following concept including discretization of the function $\underline{\psi}$ is suggested.

The function $\underline{\psi}$ is discretized in such a way that from given viewpoints - e.g.resolution requirements - N values $\underline{\nu}_i$ (i = 1...N) are selected from $\Sigma \subset \mathbb{R}^m$ and assigned to the index variable i (see Figure 4).

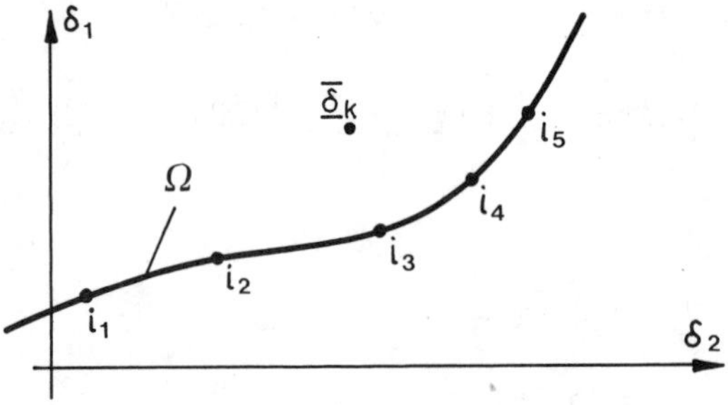

Figure 4: Discretization of $\underline{\psi}$

By means of this procedure, N values $\underline{\delta}_i$ from $\Omega \subset \mathbb{R}^q$ are also defined, which, together with $\underline{\nu}_i$ and perhaps given weighting factors $c_i \in \mathbb{R}$ (see below), are calculated and stored before the actual identification. i corresponds to the storage address. The storage requirements are N(q+m+1) cells.

From equation (4.1), for the loss function being minimized, with $(\underline{\bar{P}}_k^{\delta})^{-1} = \underline{M}_k^T \underline{M}_k$ ($\underline{M}_k$: triangular matrix) holds:

$$
\begin{aligned}
J_k &= (\underline{\psi}(\underline{\nu}) - \underline{\bar{\delta}}_k)^T (\underline{P}_k^{\delta})^{-1} (\underline{\psi}(\underline{\nu}) - \underline{\bar{\delta}}_k) - \ln g(\underline{\nu}) \\
&= (\underline{\psi}(\underline{\nu}) - \underline{\bar{\delta}}_k)^T \underline{M}_k^T \underline{M}_k (\underline{\psi}(\underline{\nu}) - \underline{\bar{\delta}}_k) - c(\underline{\nu}) \qquad (4.4) \\
&= \| \underline{M}_k (\underline{\psi}(\underline{\nu}) - \underline{\bar{\delta}}_k \|^2 - c(\underline{\nu}) \quad .
\end{aligned}
$$

From this follows, for the MAP or ML Estimation:

$$
\begin{aligned}
\hat{\underline{\nu}}_k^{MAP} &= \arg \min_{\nu \in \Sigma} \{ \| \underline{M}_k (\underline{\psi}(\underline{\nu}) - \underline{\bar{\delta}}_k) \|^2 - c(\underline{\nu}) \} \\
&\approx \underline{\nu}_{\hat{i}_k^{MAP}} \quad \text{with} \qquad (4.5)
\end{aligned}
$$

$$
\hat{i}_k^{MAP} = \arg \min_i \{ \| \underline{M}_k (\underline{\delta}_i - \underline{\bar{\delta}}_k) \|^2 - c_i \} , \qquad (4.6)
$$

or

$$\hat{\underline{\gamma}}_k^{ML} = \arg \min_{\underline{\gamma} \in \Sigma} \| \underline{M}_k \, (\underline{\psi}(\underline{\gamma}) - \bar{\underline{\delta}}_k) \|$$

$$\approx \underline{\gamma}_{\hat{i}_k^{ML}} \quad \text{with} \tag{4.7}$$

$$\hat{i}_k^{ML} = \arg \min_i \| \underline{M}_k (\underline{\delta}_i - \bar{\underline{\delta}}_k) \| \quad . \tag{4.8}$$

Carrying-out of the complete parameter estimation is done in five steps according to the following procedure:

Step 1: Discretization of $\underline{\psi}$ i.e. choice of $\underline{\gamma}_i$, N, and calculation of $\underline{\delta}_i$, c_i (this step is done off-line before the identification).

Step 2: Recursive calculation of $\bar{p}(\underline{\delta}|\underline{y}^k) = N(\bar{\underline{\delta}}_k, \bar{\underline{P}}_k^{\delta})$ from $\bar{\underline{\delta}}_{k-1}$, $\bar{\underline{P}}_{k-1}^{\delta}$ and $\underline{y}_k$. This step corresponds to the first part of the identification represented in Figure 4.

Step 3: Changing the $q \times q$ matrix $\bar{\underline{P}}_k^{\delta}$ into the triangular matrices $\underline{D}_k$, $\underline{D}_k^T$ using Cholesky's method(see equation (4.9)).

Step 4: Inversion of the triangular matrix $\underline{D}_k$ i.e. calculation of $\underline{M}_k = \underline{D}_k^{-1}$ (elimination method /13/).

Step 5: Minimization of $\| \underline{M}_k (\underline{\delta}_i - \bar{\underline{\delta}}_k) \|^2 - c_i$, i.e. determination of the value $\hat{i}$. $\hat{i}$ is the storage address of $\hat{\underline{\gamma}}_i$.

The number N of discretization points depends on the required accuracy. N may be small, if the approximation error is to be decreased by an appropriate interpolation method.

The matrix inversion in Step 4 requires only one division, but the method of Cholesky which is given by the procedure /13/

$$d_{ik} = (p_{ik} - d_{1i}d_{1k} - d_{2i}d_{2k} - \dots - d_{i-1,i}d_{i-2,k})/d_{ii},$$

$$d_{ii}^2 = p_{ii} - d_{1i}^2 - d_{2i}^2 - \dots - d_{i-1,i}^2 \tag{4.9}$$

$$\underline{D}_k = (d_{ij}) \quad , \quad \underline{\bar{P}}_k^{\delta} = (p_{ij}) \quad ,$$

requires $q^3/6$ multiplications (see step 2: $2q^2$ multiplications with a scalar system).

Steps 3 to 5 represent the second part of the identification (see Figure 3). As will be seen later, the numerical extent of these three steps can be decreased in most cases by at least 50%. Before consideration of details, a further approach is represented which doesn't require steps 3 and 4, and which allows the use of any estimation method in step 2 (part 1).

Starting with $\underline{\bar{P}}_k^{\delta} = \underline{E}$ and $g(\underline{\delta})$ = constant leads to the reduced criterion

$$J_k^E = \| \underline{\psi}(\underline{\gamma}) - \underline{\bar{\delta}}_k \| \quad ,$$

i.e. discretized

$$J_k^E = \| \underline{\delta}_i - \underline{\bar{\delta}}_k \| \quad . \tag{4.10}$$

Then the estimate

$$\underline{\hat{\gamma}}_k^E = \arg \min_{\underline{\gamma} \in \Sigma} \| \underline{\psi}(\underline{\gamma}) - \underline{\bar{\delta}}_k \| = \underline{t}^E(\underline{\bar{\delta}}_k) \approx \underline{\gamma}_{\hat{i}_k^E} \tag{4.11}$$

with

$$\hat{i}_k^E = \arg \min_{i} \| \underline{\delta}_i - \underline{\bar{\delta}}_k \| \quad ,$$

is the value $\underline{\gamma} \in \Sigma$ for which $\underline{\delta} = \underline{\psi}(\underline{\gamma})$ has the minimal Euclidean distance to $\underline{\bar{\delta}}_k$ (see Figure 5).

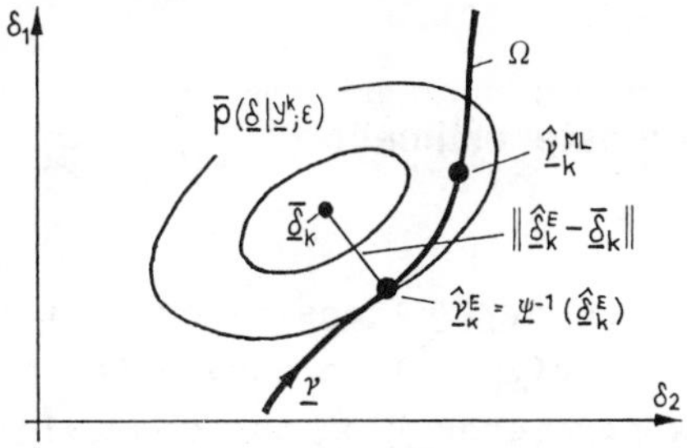

Figure 5: Representation of the reduced criterion

The parameter estimation with this modified criterion leads approximately to the MAP estimation, if the density $\bar{p}(\underline{\delta}\,|\,\underline{y}^k)$ is equally expanded in all directions with $\underline{\bar{P}}_k^{\delta} \approx c\underline{E}$, whereby c is an arbitrary constant from which the estimate and the criterion is independent. The estimation may be less effective if this symmetry is not given (e.g. fewer, minimally-disturbed observations in $\underline{y}^k$).

Besides the economizing of steps 3 and 4, the numerical extension in step 5 is essentially decreased by $\underline{M} = \underline{E}$. It can be decreased further by replacing the Euclidean norm $\|\underline{\delta}_i - \bar{\underline{\delta}}_k\| = \|\tilde{\underline{\delta}}\|_2$ with $\|\underline{\delta}_i - \bar{\underline{\delta}}_k\|_\infty = \|\tilde{\underline{\delta}}\|_\infty = \sum_{j=1}^{q} |\tilde{\delta}_j|$.

This simplified criterion is especially applicable if, as a result of the first part of the identification, not a density $\bar{p}(\underline{\delta}|\underline{y}^k)$ or its parameter $\bar{\underline{\delta}}_k$, $\underline{\bar{P}}_k^{\delta}$ is given, but only an estimate $\hat{\underline{\delta}}_k$. Hence all known on-line parameter estimation methods may be used in the first part.

A great advantage, which is the only real reason for applying the two-step concept in many cases, is the feasibility of distributing the numerical operations of part two (steps 3 to 5) over several sample intervals, which is equivalent to an estimation delay. The numerical work is decreased by the factor 1/N. In most practical problems a delay by one sample can be allowed; hence the computer load with respect to steps 3 to 5 can be reduced by at least 50%.

II.5 Combined state-parameter estimation

(a) Observer Form

The method of maximizing $p(\underline{\gamma}|\underline{Y}^k)$ considered in the last section could be transferred to $p(\underline{z}_k^*|\underline{Y}^k)$ to obtain the MAP-estimate of the state $\underline{z}_k$ and the parameter $\underline{\gamma}$ of the general system defined in equation (2.1). The numerical work would be very large, since the dimension of the argument $\underline{z}_k^*$ is greater by n (order of the system) as in the case of parameter estimation. As shown in the following, there is an easier way by calculating the state parameter estimate from the results of the parameter estimation.

At first, the state-parameter estimation in observer form is considered, by means of which not $\underline{x}_k$, $\underline{\delta}$ but $\underline{x}_k$, $\underline{\gamma}$ are to be estimated. It can be expected that the knowledge of $\underline{\delta} = \underline{\psi}(\underline{\gamma})$ will lead to improved state estimates.

When maximizing $p(\underline{x}_k, \underline{\gamma}|\underline{Y}^k)$ we must distinguish between several estimates:

$$\hat{\underline{x}}_k^{MAP} = \arg\max_{\underline{x}_k \in R^n} p(\underline{x}_k|\underline{Y}^k) \tag{5.1}$$

$$= \arg\max_{\underline{x}_k \in R^n} \int_{\Sigma} p(\underline{x}_k, \underline{\gamma}|\underline{Y}^k)\, d\underline{\gamma} \quad ,$$

$$\hat{\underline{\gamma}}_k^{MAP} = \arg\max_{\underline{\gamma} \in \Sigma} p(\underline{\gamma}|\underline{Y}^k)$$

$$= \arg\max_{\underline{\gamma} \in \Sigma} \int_{R^n} p(\underline{x}_k, \underline{\gamma}|\underline{Y}^k)\, d\underline{x}_k \quad , \tag{5.2}$$

$$\widehat{\begin{bmatrix} \underline{x}_k \\ \underline{\gamma} \end{bmatrix}}^{MAP} = \begin{bmatrix} \hat{\underline{x}}_k^{MAPV} \\ \hat{\underline{\gamma}}_k^{MAPV} \end{bmatrix} = \arg\max_{\substack{\underline{x}_k \in R^n \\ \underline{\gamma} \in \Sigma}} p(\underline{x}_k, \underline{\gamma} \,|\, \underline{Y}^k) \ . \tag{5.3}$$

The shorthand notation MAP is related to the maximum of the corresponding marginal density, while the notation MAPV refers to the maximizing of the joint density. In practical applications the difference has no essential meaning, and it doesn't exist in symmetric densities. The MAPV-estimation, however, requires much less numerical work, since there is no change-over to marginal density.

With the assumption

$$\bar{p}(\underline{x}_k^* | \underline{y}^k) = \bar{p}(\underline{x}_k, \underline{\delta} \mid \underline{y}^k) = N(\bar{\underline{x}}_k^*, \bar{\underline{P}}_k^*) ,$$

$$\bar{\underline{x}}_k^* = \begin{bmatrix} \bar{\underline{x}}_k \\ \\ \underline{\delta}_k \end{bmatrix} , \qquad \bar{\underline{P}}_k^* = \begin{bmatrix} \bar{\underline{P}}_k^x & \bar{P}_k^{x\delta} \\ \\ \bar{\underline{P}}_k^{\delta x} & \bar{\underline{P}}_k^{\delta} \end{bmatrix} ,$$

$\hat{\underline{x}}_k^{MAPV}$ and $\hat{\underline{\gamma}}_k^{MAPV}$ are given by (for derivation see Appendix):

$$\hat{\underline{\gamma}}_k^{MAPV} = \hat{\underline{\gamma}}_k^{MAP} \qquad (5.4)$$

$$\hat{\underline{x}}_k^{MAPV} = \bar{\underline{x}}_k + \underline{P}_k^{x\delta} (\underline{P}_k^{\delta})^{-1} [\underline{\psi}(\hat{\underline{\gamma}}_k^{MAP}) - \bar{\underline{\delta}}_k] \qquad (5.5)$$

$\hat{\underline{x}}_k^{ML}$ follows accordingly from equation (3.12):

$$\hat{\underline{x}}_k^{ML} = \bar{\underline{x}}_k + \underline{P}_k^{x\delta} (\underline{P}_k^{\delta})^{-1} [\underline{\psi}(\hat{\underline{\gamma}}_k^{ML}) - \bar{\underline{\delta}}_k] \qquad (5.6)$$

The numerical work required for the calculation of $\hat{\underline{x}}^{MAPV}$ is relatively small in that the parameter estimation presented in section 4 results in the product

$$\underline{s}_k = \underline{M}_k [\underline{\psi}(\hat{\underline{\gamma}}^{MAP}) - \bar{\underline{\delta}}_k] .$$

Hence, with

$$\underline{\hat{x}}_k^{MAPV} = \underline{\bar{x}}_k + \underline{\bar{P}}_k^{x\delta}(\underline{M}_k^T\underline{s}_k) = \underline{\bar{x}}_k + \Delta\underline{\hat{x}}_k \qquad (5.7)$$

only $nq + \frac{q^2}{2}$ multiplications have to be carried out ($\underline{M}_k$: triangular matrix). This is also valid for $\underline{\hat{x}}_k^{ML}$.

$\Delta\underline{\hat{x}}_k$ in equation (5.7) is that part (correction term) of the state estimate which results from the additional knowledge about $\underline{\psi}$. The greater the quotient $q/m = \dim(\underline{\delta})/\dim(\underline{\nu})$, the smaller the mean error of the improved estimate $\underline{\hat{x}}_k = \underline{\bar{x}}_k + \Delta\underline{\hat{x}}_k$ will be.

(b) Arbitrary State Representation

The MAP estimate of the state-parameter vector $\underline{z}_k^* = [\underline{z}_k^T, \underline{\nu}^T]^T$ of system (2.1) is given by(for derivation see Appendix):

$$\underline{\hat{z}}_k^{*\,MAP} = \begin{bmatrix} \underline{\hat{z}}_k^{MAPV} \\ \underline{\hat{\nu}}_k^{MAPV} \end{bmatrix} = \arg\max_{\underline{z}_k^*} p(\underline{z}_k^* \mid \underline{y}^k)$$

$$\underline{\hat{\nu}}_k^{MAPV} = \arg\max_{\underline{\nu}\in\Sigma} \left\{ \left|\det \underline{T}(\underline{\nu})\right| \cdot p(\underline{\nu} \mid \underline{y}^k) \right\} \qquad (5.8)$$

$$\underline{\hat{z}}_k^{MAPV} = \left\langle \underline{T}(\underline{\hat{\nu}}_k^{MAPV}) \right\rangle^{-1} \left\langle \underline{\bar{x}}_k + \underline{\bar{P}}_k^{x\delta}(\underline{\bar{P}}_k^{\delta})^{-1}(\underline{\psi}(\underline{\hat{\nu}}_k^{MAPV}) - \underline{\bar{\delta}}_k) \right\rangle \qquad (5.9)$$

If in a region about $\underline{\hat{\nu}}_k^{MAPV}$, in which $p(\underline{\nu} \mid \underline{y}^k)$ is essentially different from zero, $|\det \underline{T}(\underline{\nu})|$ does not depend strongly on $\underline{\nu}$, then

$$\underline{\hat{\nu}}_k^{MAPV} \approx \underline{\hat{\nu}}_k^{MAP} \qquad (5.10)$$

$$\underline{\hat{z}}_k^{MAPV} \approx \left\langle \underline{T}(\underline{\hat{\nu}}_k^{MAP}) \right\rangle^{-1} \underline{\hat{x}}_k^{MAPV} \qquad (5.11)$$

The maximum-likelihood estimate $\underline{\hat{z}}_k^{ML}$ may be easily derived by utilizing its invariance property/11/:

In general, from $\underline{z}_k^* = \underline{\psi}^{*-1}(\underline{x}_k^*)$ follows

$\underline{\hat{z}}_k^{*ML} = \underline{\psi}^{*-1}(\underline{\hat{x}}_k^{*ML})$, especially with equation (2.15):

$$\underline{\hat{z}}_k^{ML} = \langle \underline{T}(\underline{\hat{\gamma}}_k^{ML})\rangle^{-1}\, \underline{\hat{x}}_k^{ML} \quad . \tag{5.12}$$

In comparison with the estimation in observer form, the additional numerical work results in the main from the calculation of $\langle \underline{T}(\underline{\hat{\gamma}}_k)\rangle^{-1}$. Note that in most applications in which a state estimation in arbitrary coordinates is required, not all elements of $\underline{z}_k$ are of interest. Hence, not all lines of $\underline{T}_k^{-1}$ have to be known. The elements of $\underline{T}_k^{-1}$ are functions of $\underline{\hat{\gamma}}_k$. This functions can be determined approximately, before the on-line identification, as was suggested in the parameter estimation.

Now the problem of combined state-parameter estimation can be solved in seven steps, whereby the first five steps are identical with the parameter estimation represented in section 4. The two additional steps are:

Step 6: Correction of the state estimate: $\underline{\hat{x}}_k = \underline{\bar{x}}_k + \Delta\underline{x}_k$ by equation (5.7), i.e. calculation of $\underline{\hat{x}}_k^{MAPV}$ or $\underline{\hat{x}}_k^{ML}$

Step 7: Retransformation: $\hat{z}_k^j = (\underline{t}_i^j)^T \underline{\hat{x}}_k$. $(\underline{t}_i^j)^T$ defines the j-th line of the transformation matrix $(T(\underline{\hat{\gamma}}_k))^{-1}$

Finally, it should be mentioned that the concept of separated identification is also applicable to the optimal state estimation with known parameters (Kalman-Bucy Filter)by carrying out the estimation at first in observer form and subsequent retransformation. Even if all states are of interest, the numerical work will decrease, since in discrete models, which are usually obtained from continuous systems, nearly all elements in the system matrices are unknown and different from zero. For instance, a non-stationary Kalman-Bucy Filter applied to a fifth-order single-input, single-output system requires $3n^3 + 4n^2 + 5n = 500$ multiplications /10/. By using the two-stepmethod, in the worst case (full retransformation) only $3n^2 + 4n = 95$ multiplications are required (157 multiplications when additional calculation of variances is desired).

II.6 Conclusions

An on-line method for parameter estimation or for combined state and parameter estimation in linear systems has been presented. The method allows the definition of the unknown quantities in an arbitrary way, i.e. corresponding to the physical characteristics of the process to be identified - e.g. mass, temperature, position, velocity, etc.

The solution of this, in general strongly non-linear and therefore complicated estimation problem, is done in two steps. At first the unknowns to be estimated are defined in such a way that well-known or simple methods are applicable. In the second step of the solution, estimates of the quantities of primary interest are obtained by a transformation of the results from step one.

The main point of this separation is the processing of the non-linearities outside the recursion loop. This leads to fundamental advantages (see section 3).

The advantage of the method depends in a high degree on the particular character of the problem statement, hence a general criticism is difficult. Nevertheless, comparisons to other on-line methods are possible, in principle. In the main, other methods approximate the non-linearities, which must be sufficiently well differentiable, by a finite Taylor-series of the first or second order. In the first place, the Extended Kalman Filter (EKF) has to be mentioned /8 /, /3 /, which is characterized by linearizing the non-linear system equations around the current estimate. Such "local" methods work badly if the estimation error is great, especially where there is a high initial uncertainty. Moreover, linearizing errors are transmitted from one sample interval to the other inside the recursion loop. Hence, assertions about error characteristics and divergence probabilities are scarcely possible. The numerical work of these methods can be very great. As an example, the EKF, applied to state-parameter estimation of an n-th order single-input, single-output system with m unknown parameters, requires the realization of, in general, $n^2+mn+m+n$ functions (nm of them have an argument with dimension(m+n)). In comparison, the two-step concept requires only (p+2)n functions with m-dimensional argument, whereby p is the number of states which have to be transformed back.

ACKNOWLEDGEMENT

This contribution (part I and II) was written within the scope of a science project sponsored by the Deutschen Forschungsgemeinschaft. I would like to thank Prof.Dr.-Ing. Ch.Landgraf for his many valuable suggestions.

II.7 Appendix

a) Derivation of the equations (3.10) to (3.13)

Equation (3.10) followes from

$$p(\underline{x}_k,\underline{\gamma}|\underline{y}^k) = c\ p(\underline{y}^k|\ \underline{x}_k,\underline{\gamma})p(\underline{x}_k,\underline{\gamma})$$

$$= c\ p_{y|x,\delta}(\underline{y}^k|\underline{x}_k,\underline{\psi}(\underline{\gamma}))p(\underline{x}_k|\underline{\gamma})p(\underline{\gamma})$$

$$= c\ p_{y|x,\delta}(\underline{y}^k|\ \underline{x}_k,\underline{\psi}(\underline{\gamma}))p_{x|\delta}\ (\underline{x}_k\ |\ \underline{\psi}(\underline{\gamma}))\bar{p}_\delta(\underline{\psi}(\underline{\gamma}))\ \frac{p(\underline{\gamma})}{\bar{p}_\delta(\underline{\psi}(\underline{\gamma}))}$$

$$= c\ p_{y|x,\delta}(\underline{y}^k|\underline{x}_k,\underline{\psi}(\underline{\gamma}))\bar{p}_{x,\delta}(\underline{x}_k,\underline{\psi}(\underline{\gamma}))\ \frac{p(\underline{\gamma})}{\bar{p}_\delta(\underline{\psi}(\underline{\gamma}))}$$

or

$$p(\underline{x}_k,\underline{\gamma}|\underline{y}^k) = c\ \bar{p}_{x,\delta|y}(\underline{x}_k,\underline{\psi}(\underline{\gamma})\ |\ \underline{y}^k)g(\underline{\gamma})\ . \tag{3.10}$$

It is furthermore true for a sufficiently small ϵ :

$$p(\underline{y}^k\ |\ \underline{z}_k^*) = p_{y|x*}(\underline{y}^k|\ \underline{\psi}^*(\underline{z}_k^*))$$

$$\approx c\ p_{y|x*}(\underline{y}^k|\ \underline{\psi}^*(\underline{z}_k^*))\bar{p}_{x*}(\underline{\psi}^*(\underline{z}_k^*)\ ;\ \epsilon)$$

$$= c\ \bar{p}_{x*|\ y}(\underline{\psi}^*(\underline{z}_k^*)\ |\ \underline{y}^k\ ;\ \epsilon) \tag{3.11}$$

with $\bar{p}_{x*}(\underline{\psi}^*(\underline{z}_k^*)\ ;\epsilon) = N(\underline{O},\ \frac{1}{\epsilon}\ \underline{E})$

or for the special case $\underline{z}_k^* = \underline{x}_k^*$:

$$p(\underline{y}^k\ |\ \underline{x}_k^*) \approx c\ \bar{p}_{x*|\ y}(\underline{x}_k^*|\underline{y}^k;\ \epsilon)\ . \tag{3.12}$$

In the same way we can calculate $p(\underline{y}^k | \underline{\gamma})$:

$$p(\underline{y}^k | \underline{\gamma}) = p_{y|\delta}(\underline{y}^k | \underline{\psi}(\underline{\gamma}))$$

$$\approx c \; p_{y|\delta}(\underline{y}^k | \underline{\psi}(\underline{\gamma})) \; \bar{p}_\delta(\underline{\psi}(\underline{\gamma}); \epsilon)$$

$$= c \; \bar{p}_{\delta|y}(\underline{\psi}(\underline{\gamma}) | \underline{y}^k; \epsilon) \qquad (3.13)$$

with

$$\bar{p}_\delta(\underline{\psi}(\underline{\gamma}); \epsilon) = N(\underline{O}, \tfrac{1}{\epsilon}\underline{E}) \quad .$$

b) Derivation of the equations (5.4) and (5.5).

With equation (3.5) we obtain

$$p(\underline{x}_k, \underline{\gamma} | \underline{y}^k) = p_{x|\delta,y}(\underline{x}_k | \underline{\gamma}, \underline{y}^k) \; p(\underline{\gamma} | \underline{y}^k)$$

$$= p_{x|\delta,y}(\underline{x}_k | \underline{\psi}(\underline{\gamma}), \underline{y}^k) \; p(\underline{\gamma} | \underline{y}^k) \; . \qquad (7.1)$$

The factor $p(\underline{x}_k | \underline{\delta}, \underline{y}^k)$ can be calculated from $\bar{p}(\underline{x}_k^* | \underline{y}^k)$. First, we get

$$\bar{p}(\underline{x}_k^* | \underline{y}^k) = \bar{p}(\underline{x}_k, \underline{\delta} | \underline{y}^k)$$

$$= c \; p(\underline{y}^k | \underline{x}_k, \underline{\delta}) \; p(\underline{x}_k | \underline{\delta}) \; \bar{p}(\underline{\delta}) \qquad (7.2)$$

with

$$c = \frac{1}{\bar{p}(\underline{y}^k)} \quad .$$

The product $p(\underline{y}^k | \underline{x}_k, \underline{\delta}) \; p(\underline{x}_k | \underline{\delta})$ can be remodeled by known rules as

$$p(y|x,\delta) \; p(x|\delta) = p(x|\delta,y) \; p(y|\delta).$$

Hence, equation (7.2) can be written as follows:

$$\bar{p}(\underline{x}_k,\underline{\delta}\ \underline{y}^k) = c\ p(\underline{x}_k|\underline{\delta},\underline{y}^k)p(\underline{y}^k|\underline{\delta})\bar{p}(\underline{\delta})$$

$$= p(\underline{x}_k|\underline{\delta},\underline{y}^k)\bar{p}(\underline{\delta}|\underline{y}^k) \quad .$$

Consequently:

$$p(\underline{x}_k|\underline{\delta},\underline{y}^k) = \bar{p}(\underline{x}_k|\underline{\delta},\underline{y}^k)$$

$$= \frac{\bar{p}(\underline{x}_k,\underline{\delta}|\underline{y}^k)}{\bar{p}(\underline{\delta}|\underline{y}^k)} = \frac{\bar{p}(\underline{x}_k,\underline{\delta}|\underline{y}^k)}{\int \bar{p}(\underline{x}_k,\underline{\delta}|\underline{y}^k)\ d\underline{x}_k}$$

with $\bar{p}(\underline{\delta}|\underline{y}^k) \neq 0$.

With the assumption

$$\bar{p}(\underline{x}_k,\underline{\delta}|\underline{y}^k) = \bar{p}(\underline{x}_k^*|\underline{y}^k) = N(\bar{\underline{x}}_k^*,\bar{\underline{P}}_k^*) ,$$

$$\bar{\underline{x}}_k^* = \begin{bmatrix} \bar{\underline{x}}_k \\ \bar{\underline{\delta}}_k \end{bmatrix} , \quad \bar{\underline{P}}_k^* = \begin{bmatrix} \bar{\underline{P}}^x & \bar{\underline{P}}^{x\delta} \\ \bar{\underline{P}}_k^{\delta x} & \bar{\underline{P}}_k^{\delta} \end{bmatrix} , \tag{7.3}$$

it follows according to the usual relationships for two random vectors with a common Gaussian distribution (see for example /8/, Theorem 2.13):

$$p(\underline{x}_k|\underline{\delta},\underline{y}^k) = \bar{p}(\underline{x}_k|\underline{\delta},\underline{y}^k)$$

$$= N\left\langle \bar{\underline{x}}_k + \bar{\underline{P}}_k^{x\delta}\ (\bar{\underline{P}}_k^{\delta})^{-1}\ (\underline{\delta}-\bar{\underline{\delta}}_k)\ ,\ \bar{\underline{P}}_k^{x} - \bar{\underline{P}}_k^{x\delta}\ (\bar{\underline{P}}_k^{\delta})^{-1}\ \bar{\underline{P}}_k^{\delta x} \right\rangle . \tag{7.4}$$

It is thus relatively simple (compare equation 7.1) to maximize the density

$$p(\underline{x}_k,\underline{\gamma}|\underline{y}^k) = p_{x|\gamma,y}(x_k|\underline{\gamma},\underline{y}^k)p(\underline{\gamma}|\underline{y}^k)$$

$$= \bar{p}_{x|\delta,y}(\underline{x}_k|\underline{\psi}(\underline{\gamma}),\underline{y}^k)p(\underline{\gamma}|\underline{y}^k) \tag{7.5}$$

Since the greatest values for the factors $p(\underline{x}_k|\underline{\delta},\underline{y}^k)$ and $p(\underline{\gamma}|\underline{y}^k)$ can be found independently of one another (the upper value of $p(\underline{x}_k|\underline{\delta},\underline{y}^k)$ is, according to equation (7.4) independent of $\underline{\delta}$ as well as $\underline{\gamma}$), we obtain

$$\arg\max_{\underline{x}_k \in R^n} p_{x|\delta,y}(\underline{x}_k|\underline{\psi}(\underline{\gamma}),\underline{y}^k) = \bar{\underline{x}}_k + \bar{\underline{P}}_k^{x\delta}(\bar{\underline{P}}_k^{\delta})^{-1}(\underline{\psi}(\underline{\gamma})-\bar{\underline{\delta}}_k)$$

$$\arg\max_{\gamma\in\Sigma} p(\underline{\gamma}|\underline{y}^k) = \hat{\underline{\gamma}}_k^{MAP} ,$$

which finally leads to the result in equations (5.4) and (5.5).

c) Derivation of the equations (5.8) and (5.9)

First, we get

$$p(\underline{z}_k^*|\underline{y}^k) = p(\underline{z}_k,\underline{\gamma}|\underline{y}^k)$$

$$= p(\underline{z}_k|\underline{\gamma},\underline{y}^k)p(\underline{\gamma}|\underline{y}^k)$$

$$= p_{x|\gamma,y}(\underline{T}(\underline{\gamma})\underline{z}_k|\underline{\gamma},\underline{y}^k)\ |\det\underline{T}(\underline{\gamma})|\ p(\underline{\gamma}|\underline{y}^k)$$

$$= p_{x|\delta,y}(\underline{T}(\underline{\gamma})\underline{z}_k|\underline{\psi}(\underline{\gamma}),\underline{y}^k)p^T(\underline{\gamma}|\underline{y}^k)$$

with $$p^T(\underline{\gamma}|\underline{y}^k) := |\det\underline{T}(\underline{\gamma})|\ p(\underline{\gamma}|\underline{y}^k) .$$

Continuing with equation (7.4):

$$p(\underline{z}^*|\underline{y}^k) = \bar{p}_{x|\delta,y}(\underline{T}(\underline{\gamma})\underline{z}_k|\underline{\psi}(\underline{\gamma}),\underline{y}^k)p^T(\underline{\gamma}|\underline{y}^k) .$$

This equation is of the same type as equation (7.5). $p(\underline{z}_k,\underline{\gamma}|\underline{y}^k)$ can be maximized in a similar manner, so that the equations (5.8) and (5.9) finally result.

II.8 References

/1/ Aoki, M., P.C. Yue — On Certain Convergence Questions in System Identification
SIAM J. Control 2 May 1975, P.139-156

/2/ Aström, K.T., Bohlin, T., Wensmark, S. — Automatic construction of linear stochastic dynamic models for stationary industrial processes with random disturbances using operating records.
Rep.TP 18.150, IMB Nordic Laboratory, Lindingö, Sweden 1965

/3/ Brammer, K. — Schätzung von Parametern und Zustandsvariablen linearer Regelstrecken durch nichtlineare Filterung.
Regelungstechnik 6/1970

/4/ Bühler, E. — A Generalization of the Kalman-Filter-Algorithm and its Application in Optimal Optimal Identification
Advances in Control Systems and Signal Processing(this volume)

/5/ Glover, K., Willems, J.C. — Parameterizations of Linear Dynamical Systems: Canonical Forms and Idetifiability
IEEE Trans on autom. control No.6/1974

/6/ Gnedenko, B.W. — Lehrbuch der Wahrscheinlichkeitsrechnung. Akademie-Verlag, Berlin 1968

/7/ Hartmann, I. — Lineare Systeme.
Springer, Berlin 1976

/8/ Jazwinski, A.H. — Stochastic Processes and Filtering Theory
Academic Press, New York, 1970

/9/ Landgraf, Ch. — Stochastische Lineare Systeme
Brennpunkt Kybernetik, TU Berlin 1976

/10/ Mendel, J. — Computational Requirements for a Discrete Kalman Filter
IDEE Trans on autom. control No.6/1971

/11/ Mood, A.M., Graybill, F.A. — Introduction to the Theory of Statistics. 3rd Edition.
New York, McGraw-Hill 1963

/12/ Staley, R.M., Yue, P.C. — On System Parameter Identifiability
Information Science 2/1970, p.127-138

/13/ Zurmühl, R. — Matrizen
Berlin, Springer 1964